TCP Performance over UMTS-HSDPA Systems

TCP Performance over UMTS-HSDPA Systems

Mohamad Assaad
Djamal Zeghlache

CRC Press
Taylor & Francis Group
Boca Raton London New York

CRC Press is an imprint of the
Taylor & Francis Group, an **informa** business
AN AUERBACH BOOK

CRC Press
Taylor & Francis Group
6000 Broken Sound Parkway NW, Suite 300
Boca Raton, FL 33487-2742

First issued in paperback 2019

ISBN-13: 978-0-8493-6838-7 (hbk)
ISBN-13: 978-0-367-39056-3 (pbk)

Library of Congress Cataloging-in-Publication Data

Assaad, Mohamad.
   TCP performance over UMTS-HSDPA systems / Mohamad Assaad and Djamal Zeghlache.
      p. cm.
   Includes bibliographical references and index.
   ISBN 0-8493-6838-3 (alk. paper)
   1. Universal Mobile Telecommunications System. 2. TCP/IP (Computer network protocol) I. Zeghlache, Djamal. II. Title.

TK5103.4883.A77 2006
621.382'12--dc22   2006042662

Visit the Taylor & Francis Web site at
http://www.taylorandfrancis.com

and the CRC Press Web site at
http://www.crcpress.com

Dedication

Dedicated to my family and friends.

M.A.

Dedicated to my family and friends and students
I have the privilege of working with.

D.Z.

Contents

The Authors

Dr. Mohamad Assaad received a B.S. in electrical engineering with highest honors from Lebanese University in Beirut, in 2001. In 2002 and 2006, he received a M.Sc. degree with highest honors and a Ph.D., both in telecommunications, from the Ecole Nationale Supérieure des Télécommunications (ENST) in Paris, France. While pursuing his Ph.D., he was a research assistant in the wireless networks & multimedia services department at the Institut National des Télécommunications (INT) in Evry, France, working on cross-layer design in UMTS/HSDPA system and interaction of TCP with MAC/RLC and physical layers. He has published several international journal and conference papers in this area and worked in collaboration with some academic and industrial partners. Since April 2006, he is assistant professor in the telecommunications department at the Ecole Supérieure d'Electricité (Supélec) in Gif-sur-Yvette, France. His research interests include 3G systems and beyond, TCP protocol in wireless networks, cross-layer design, resource allocation in wireless systems, multiuser detection, and MIMO techniques.

Dr. Djamal Zeghlache graduated from Southern Methodist University in Dallas, Texas, in 1987 with a Ph.D. in electrical engineering and the same year joined Cleveland State University as assistant professor in the area of digital communications in the Department of Electrical Engineering. During the summer of 1990 and 1991 he was a summer faculty fellow at the NASA Lewis Research Center where he conducted research and development on a portable terminal for satellite communications in the Ka Band. In 1992 he joined the Institut National des Télécommunications in Evry, France, where he currently heads the Wireless Networks & Multimedia Services Department. He is actively involved in European funded projects, is a member of IEEE Technical Committee on Personal Communications of the IEEE Communications Society, and a member of the Wireless World Research Forum. He has coauthored papers on multiple access, wireless resource management, wireless network planning, and networking for personal networks. He also regularly partakes in the technical management of most IEEE conferences as member of technical program committees and as reviewer in conferences such as ICC, Globecom, PIMRC, ASWN, WPMC, and WCNC, to name a few. He was cotechnical chair of the wireless communications symposium of Globecom 2003 and cofounder of the ASWN workshop that he cochaired in 2001, 2002, and 2005. Dr. Zeghlache's research interests and activities span a broad spectrum of issues related to wireless networks and services including radio access (cross-layer design, resource management, and planning) and core networks with emphasis on end-to-end issues such as cooperation between networks and dynamic adaptation of context aware wireless networks and services.

Preface

Wireless systems and networks gradually evolved from voice centric first-generation technologies to digital systems offering in addition non real-time low data rate services. Despite this evolution from first to second generation, data rates for cellular systems have remained fairly low. Wireless local area networks on the contrary offer higher aggregate data rates and are, from conception, already compatible with the Internet and its associated protocols. The introduction of packet-oriented cellular networks that are Internet Protocol (IP) compatible or can easily interwork with the Internet appeared in the late Nineties with GPRS, EDGE, IS-95, and IS-136 systems. This has been motivated by the need for cellular networks and services to become compatible with the Internet and related multimedia services to achieve converge. Without this merging, cellular networks will not benefit from the tremendous Internet growth and related advances in multimedia applications and services.

The increased use of Internet and data services also motivated the introduction of packet-oriented systems. Various advanced radio technologies can also facilitate the introduction of mass-market multimedia services in wireless networks. The most frequently used techniques are adaptive modulation and coding (AMC) to achieve better spectral efficiency, link adaptation to mitigate radio channel impairments, scheduling to enable intelligent allocation and sharing of resources to improve capacity, and sophisticated detection and decoding techniques to handle multiuser interference, to name a few approaches improving wireless networks. Multiple transmit and receive antennas can also be used to achieve higher data rates and to improve system capacity. In fact, achieving high data rates requires the introduction of space, time, and frequency diversity techniques. Wireless local area networks have started using diversity already. Cellular systems are expected to take full advantage of these three dimensions in the near future. For example the Universal Mobile for Telecommunications System (UMTS) technology in Europe is preparing for this evolution and convergence in successive releases within the Third-Generation Partnership Project (3GPP) that standardizes and specifies the UMTS air interface and the access and core network architectures. The introduction of multiple antennas is planned for the last phases of the UMTS architecture enhancements.

Wireless systems, whether TDMA, CDMA, or OFDM based, have so far been very timid at introducing these essential features. Third-generation cellular is expected to introduce AMC, scheduling, and diversity techniques to improve spectral efficiency (capacity per cell) and data rates (per session or application) in multiple phases. The UMTS FDD mode, corresponding to the WCDMA standard in Europe, will include AMC, scheduling, and hybrid ARQ starting from Release 5 and beyond.

However, Release 99 of UMTS, the first release, relies mostly on the introduction of CDMA-based radio and access technologies. This release uses for its packet domain the GPRS core network to provide partial IP convergence through tunneling protocols and gateways. This step in the evolution remains insufficient to achieve full compatibility with IP. The air interface as specified in Release 99 does not provide the needed higher data rates either. Beyond this release, meant to provide smooth transition from GSM and GRPS to UMTS 3G, Releases 5 through 7 introduce a number of additional enhancements into the standard to enable flexible and adaptive packet transmissions and to offer Internet-based services. The Session Initiation Protocol (SIP) from the Internet Engineering Task Force (IETF) was also adopted by 3GPP for UMTS to establish and control sessions just like on the Internet. This has led to the introduction of an Internet Multimedia Subsystem (IMS) to strengthen convergence.

At the data link layer (Radio Link Control and Medium Access Control) and the radio resource control, the first major enhancement was the addition in the downlink of shared channels next to the Release 99 dedicated channels. The dedicated channels are suitable for real-time services but are inadequate for packet services. Precious resource (corresponding to power and codes in CDMA) would be wasted and much capacity lost if only dedicated channels were used. The introduction of shared channels results in power savings, interference mitigation, and system capacity improvements. More recently, enhancements have been added in the uplink direction as well.

As mentioned previously, enhancing data rates can be achieved by introducing Adaptive Modulation and Coding associated with radio link adaptation to channel conditions. Most current systems integrate AMC and introduce additional techniques to enhance data rates and reliability over the air. The UMTS data link layer uses HARQ to retransmit erroneously received radio blocs and to increase link reliability. Standardized measurements and quality indicators in UMTS provide the means to achieve efficient modulation and coding selection and link adaptation.

Besides the introduction of AMC in the standard, scheduling over the shared channels is added to improve capacity and to offer packet-based multimedia services. Scheduling over shared channels must take into account radio channel conditions, mobile location in the cell, and service type to provide tangible throughput, capacity, and delay benefits. Scheduling must also ensure fairness with respect to users and applications.

The introduction of new features in networks to improve data rates and to enhance reliability of data transmission over the air interface nevertheless can have an impact on end-to-end performance and efficiency. Retransmission mechanisms relying on ARQ interact with higher-layer protocols, especially the Transport Control Protocol (TCP) used in conjunction with IP to offer nonreal-time services. Real-time services are typically offered using User Datagram Protocol (UDP)/IP and streaming services using Real-Time Streaming Protocol (RTSP)/Real-Time Transport Protocol (RTP)/IP. Cross-layer interactions can have a drastic impact on overall throughput and capacity. Care must be taken to characterize these interactions and to suggest ways of preventing or at least reducing any negative effects resulting from the introduction of ARQ and other techniques in wireless networks that unavoidably interact with congestion control mechanisms in core networks.

The interaction between radio link control mechanisms and TCP was identified early in the scientific community and has since provided many variants for TCP to reduce and possibly to eliminate interactions when random errors over the air interface are mistakenly taken by TCP as congestion in the fixed network segments. The plethora of TCP variants available today are described in Chapter 6 of this book. Even if some approaches propose link-layer solutions, most tend to break the end-to-end IP paradigm when TCP is modified in an attempt to alleviate the experienced negative cross-layer effects due to errors occurring over the radio link. Among the proposed solutions, only a few are actually used in practice. Split TCP has been used in public land mobile networks (PLMNs) at gateways located at the edge of wireless core networks to separate the Internet from the PLMNs, thereby avoiding interactions between TCP and the radio link errors and recovery mechanisms. Some TCP versions have also become de facto standards because they have been extensively deployed in the Internet during the quest for alternatives to the standard or the original TCP. This book consequently focuses in Chapter 7 on the more common and popular versions of TCP when conducting an analysis for UMTS using HARQ and scheduling.

This book has been structured to enable readers already familiar with either UMTS or TCP and its variants to skip certain chapters and move directly to parts of the manuscript of main interest to them. The actual structure of the book is depicted in Figure 1. There are two main parts presented in the manuscript. The first, Chapters 1 through 4, provides background and

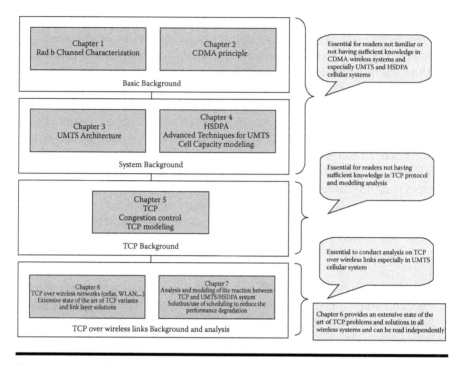

Essential for readers not familiar or not having sufficient knowledge in CDMA wireless systems and especially UMTS and HSDPA cellular systems

Essential for readers not having sufficient knowledge in TCP protocol and modeling analysis

Essential to conduct analysis on TCP over wireless links especially in UMTS cellular system

Chapter 6 provides an extensive state of the art of TCP problems and solutions in all wireless systems and can be read independently

Figure 1   General structure of the book.

state-of-the-art for wireless networks with some emphasis on the UMTS network as one of the third-generation wireless technologies. Also included in this part is analysis and modeling of the overall cell capacity for both UMTS Release 99 and high-speed downlink packet access (HSDPA) systems. The second part, found in Chapters 5 through 7, focuses on the interaction of TCP with wireless systems. It provides an extensive state-of-the-art of TCP over wireless networks and conducts analysis and mathematical modeling for the interaction of HARQ and TCP in UMTS networks including HSDPA capability.

Chapters 1 through 4 provide background information on the radio propagation channel and the typical network architectures for cellular with emphasis on GSM and GPRS to build understanding for the transition to UMTS networks. UMTS networks are characterized later in terms of their architecture and advanced features such as AMC, HARQ, and scheduling. Chapter 1 is a summary of the radio propagation channel in terms of path loss, fast fading, and shadowing. The behavior of the channel response is also reported along with a section describing propagation reference models specified in the UMTS standard to design systems for various environments. Chapter 2 covers briefly the CDMA principle to prepare readers, not familiar with CDMA systems, for Chapter 3 where the UMTS network is first introduced.

Chapter 3 presents the general architecture for UMTS, followed by a description of physical and logical channels as defined in the standard. The UMTS radio link layer and medium-access control along with power control and handover aspects complete the chapter.

Chapter 4 describes the HSDPA introduced in Release 5 of UMTS. HSDPA enhances data rates by introducing AMC, HARQ, and scheduling over the shared channels that are extended with high-speed scheduling capability. A new timing structure with finer-grain transmission intervals as low as 2 ms is also created to achieve efficient scheduling and to obtain meaningful capacity gains. Chapter 4 also provides a statistical characterization of the number of retransmissions when using HARQ based on chase combining. This result, along with scheduling and cell capacity models of Chapter 4, serves as background to derive UMTS HSDPA cell capacity for end-to-end communications later on in Chapter 7.

Chapter 5 is devoted to the TCP used in the Internet and serves as background for Chapters 6 and 7, which address TCP over wireless and the cross-layer interactions encountered when radio link control mechanisms and TCP are used in the end-to-end path between hosts. Chapter 6 is an extensive state-of-the-art of TCP over wireless networks. The chapter presents an exhaustive list of TCP versions and link layer solutions that consist of adapting TCP (often modifying the original TCP) to wireless networks. TCP was designed for handling congestion in fixed networks and the Internet and is not suitable as is for wireless communications via interconnecting infrastructures.

Chapter 7 appears at first glance as an analysis of interactions between Radio Link Control (RLC) and TCP specialized to UMTS HSDPA. In fact, this chapter goes further, as it clearly contends that systems using scheduling over the air interface are better off taking advantage of scheduling to alleviate, if not eliminate, the RLC–TCP interactions rather than violating the end-to-end IP paradigm. An analytical model is proposed to assess the cell throughput for HSDPA using proportional fair scheduling and the de facto TCP Reno congestion control algorithm. The results reported in Chapter 7 indicate that wireless systems can rely on scheduling to minimize interactions of lower-layer protocols with TCP and can achieve capacity benefits without breaking the IP paradigm. An analytical expression of cell throughput provides insight on capacity behavior. This mathematical model can be used to assess performance trends in UMTS releases using HSDPA technology.

Figure 1 is meant to provide potential hints on prerequisites for each chapter and to indicate the chapters that can be skipped by readers familiar with wireless systems or Internet protocols but not both. Readers with knowledge in wireless systems and CDMA principles can skip Chapters 1 and 2. Chapter 3, providing background knowledge on the UMTS network architecture, is of interest to people not familiar with third-generation wireless. These three chapters and Chapter 4 are essential for those who have only knowledge on the Internet and its protocols. Chapter 4 must be read by all persons not familiar with HSDPA. Chapter 5 describes the TCP and provides the background needed for Chapters 6 and 7. Chapter 5 can be skipped by readers with good knowledge of the IP and TCP. Chapters 6 and 7 address TCP performance and usage over wireless networks with Chapter 7 conducting the analysis, the modeling, and the performance study for the UMTS HSDPA system. It explores the use of scheduling to reduce interactions between TCP and the HSDPA radio link layer mechanisms.

Chapter 1

Wireless Radio Channel*

Wireless signals transmitted over the wireless propagation channel are subject to a number of impairments and effects that must be characterized to ease the conception and design of wireless systems—especially antenna systems and receivers. Since the radio propagation channel has been widely covered and characterized in the literature, this book limits the description to an overview and only emphasizes fundamentals and aspects mostly relevant to system design and capacity estimation. In system design, the objective is to maximize capacity and operation under various propagation conditions. For HSDPA, the requirement is to achieve this optimization for urban environments.

Received signals in wireless are subject to a number of phenomena including

- Path loss due to distance between the transmitter and receiver. Shadowing effects due to the immediate environment around the user that influence the average received signal energy over several hundreds of wavelengths.
- Fast fading in the signal envelope manifest via random signal amplitude, phase, energy, and power variations. These variations are observed on timescales of the order of half a wavelength.
- Doppler effects due to user or terminal mobility. This is a frequency-spreading phenomenon proportional to vehicle or terminal speed.

These effects result from wave propagation through the wireless channel. The waves are subject to reflection, diffraction, and scattering over obstacles.

*This chapter was coauthored by Abdelwaheb Marzouki, assistant professor at the Institut National des Telecommunication, Evry, France.

Reflection occurs when waves impinge on objects of large dimensions when compared with the wavelength of the propagation wave. Reflection depends on material properties, wave polarization, angle of incidence, and operation frequency. Diffraction occurs when the path between the transmitter and the receiver is obstructed by a surface with sharp irregularities (edges), like buildings. When a single object causes shadowing, the attenuation can be estimated by treating the obstacle as a diffracting knife edge. Scattering occurs when the medium through which the wave travels is composed of objects with small dimensions, when compared to the wavelength, and where the number of obstacles is large. Scattered waves are produced when waves impinge in rough surfaces, foliage, and small objects in general. Terrain irregularities, land type, and environment morphology also influence wireless signals. These effects are typically included in propagation models and are taken into account in network planning.

When modeling the propagation channel, simplifying assumptions give rise to expressions of signal energy or power at a receiver antenna as a function of transmitter to receiver distance, frequency, antenna heights, and environment type—rural, suburban, urban, and indoor. This is often sufficient to conduct initial and approximate network planning.

The design of receivers, especially advanced receivers, requires precise characterization of the received signal. The propagation channel is viewed in this case as a nonlinear and time-varying filter. Models focus on the impulse response characterization and the statistics of the signal envelope and phase.

In summary, there are four aspects to analyze and characterize when studying the mobile propagation channel: path loss, shadowing, fast fading, and the doppler effect. The fading aspects can be further classified into large-scale and small-scale fading as depicted in Figure 1.1. Large-scale fading corresponds to average signal power attenuation due to motion over large areas. Small-scale fading represents the variation of the attenuation

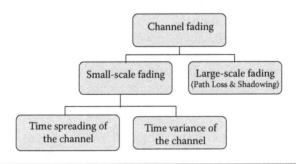

Figure 1.1   Classification of channel fading aspects (large and small scale fading).

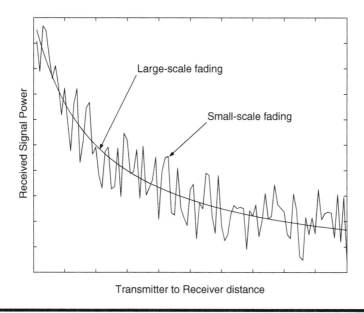

Figure 1.2   Effect of large and small scale fading of the radio channel on the received signal power.

over few wavelengths and is caused by multipath propagation. Due to reflection over obstacles, multiple copies of the transmitted signal are received at the antenna, which gives rise to a composite signal with rapid variations in amplitude and phase. This is known as fast fading or small-scale fading since it occurs over timescales of the order of a wavelength (see Figure 1.2).

Fast fading due to multipath is characterized by analyzing the power delay profile of the received multiple paths or signal copies. The power delay profile informs on the delay spread between the first and last path received and on the energy or power available in each path. This time spreading of the signal energy over time is known as the delay spread. Terminal or user mobility and speed cause Doppler spreading of the carrier frequency or the system operating frequency. By time-to-frequency duality through the Fourier transform, it is possible to show that the propagation channel acts as a double spread or dispersive channel caused by both the multipath phenomenon and the user or terminal velocity. Multipath induces frequency selective fading, and the Doppler spreading causes the time selective nature of the mobile radio propagation channel.

Now some detailed characterization of these propagation channel phenomena are provided, starting with large-scale fading.

1.1 Large-Scale Fading Models

Large-scale fading describes the variation of the mean path loss over long time periods. It is a random variable, with a log normal distributed mean, given by the relation

$$L_p(d) = X_a + \overline{L_p(d)}, \tag{1.1}$$

where

- d is the distance between the transmitter and receiver;
- X_a is a zero mean Gaussian random variable expressed in decibels;
- $\overline{L_p(d)}$ is the average path loss, over a multitude of different sites, for a given value of d.

The log normal distribution has been obtained by statistical fitting, and the expression of the mean path loss is given by

$$f_A(a) = \frac{1}{\sqrt{2\pi}\sigma a} e^{-\frac{(\log(a) - \overline{L_p(d)})^2}{2\sigma^2}}, \quad a > 0. \tag{1.2}$$

It is sometimes useful, especially when dealing with cell planning, to write the mean path loss in the following form:

$$\overline{L_p(d)}(dB) = \overline{L_s(d_0)}(dB) + 10n \log(d/d_0), \tag{1.3}$$

where $\overline{L_s(d_0)}$ is path loss of the signal at a reference distance d_0. The reference distance d_0 corresponds to a point located in the far field of the antenna. Typically, the value of d_0 is taken to be 1 km for large cells, 100 m for microcells, and 1 m for indoor channels. The value of n depends on the type of environment. For free space $n = 2$, in presence of strong guided waves $n < 2$ and $n > 2$ when huge obstacles are present. The coefficient n may also be obtained by equating (1.3) with the path-loss model of the channel. The latter is obtained through a combination of experimental measurement fittings and physical assumptions.

1.1.1 Path Loss Models for UMTS

Several path loss models for cellular systems were proposed by the International Telecommunications Union (ITU), ETSI, and 3GPP. Since the present book concerns UMTS HSDPA systems, three models describing path loss in environments where UMTS HSDPA is likely to be deployed are of direct interest [1]. These correspond to the following environments: indoor office, outdoor-to-indoor pedestrian, and vehicular.

1.1.1.1 Path Loss Model for Indoor Office Environment

Office environments are served by pico-cells, with a radius less than 100 m, characterized by low transmit powers. The base stations and the pedestrian users are located indoor. The path-loss variation is essentially due to scatter and attenuation by walls, floors, and metallic structures. These objects also produce shadowing effects. A log-normal shadow fading standard deviation of 12 dB can be expected.

One path loss model for this environment is based on the COST 231 [2] model defined as follows:

$$\overline{L_p(d)} = L_{FS} + L_c + \sum_i k_{wi} L_{wi} + n^{((n+2)(n+1)-b)} L_f, \qquad (1.4)$$

where

- L_{FS} = free space between transmitter and receiver
- L_c = constant loss, typically set at 37 dB
- k_{wi} = number of penetrated walls of type i
- n = number of penetrated floors; $n = 4$ is an average for indoor office environments.
- L_{wi} = loss of wall type i; typical values are $L_{w1} = 3, 4$ for plasterboard and windows and $L_{w2} = 6, 9$ for brick.
- L_f = loss between adjacent floors; a typical value of L_f is 18.3 dB
- b = empirical parameter with typical value of 0.46

1.1.1.2 Path Loss Model for Urban and Suburban Environment

Urban and suburban environments are characterized by large cells and high transmit powers with a standard deviation around 10 dB. A path loss used for the test scenarios in urban and suburban areas is described in [1]. This model, called vehicular model, is derived from the COST 231 [2] outdoor model and has been adopted by ITU. Despite being called vehicular this model has nothing to do with vehicles. The path loss in the vehicular model is given by

$$\overline{L_p(d)} = -20 \log\left(\frac{\lambda}{4\pi d}\right) - 10 \log\left(\frac{\lambda}{2\pi^2 r}\left(\frac{1}{\theta} - \frac{1}{2\pi + \theta}\right)^2\right)$$

$$-10 \log\left(2.35^2 \left(\frac{\Delta h_b}{d}\sqrt{\frac{l}{\lambda}}\right)^{1.8}\right) \qquad (1.5)$$

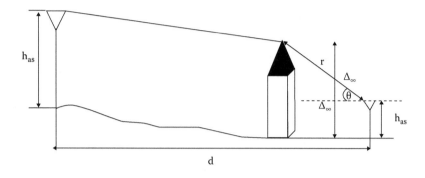

Figure 1.3 Vehicular channel model adopted by ITU [1] for urban and suburban area.

with parameters specified following (see Figure 1.3):

- b_{BS} is base station antenna height (in m)
- b_{MS} is mobile station antenna height (in m)
- Δm is mean building rooftop height (in m)
- $\Delta b_b = b_{BS} - \Delta m$ is difference between the base station antenna height and mean building rooftop height (in m). It ranges from 0 to 50 m; typically $\Delta b_b = 15$ m.
- l is average separation between the buildings (in m)
- Δmr is mean building height (in m)
- θ is the angle subtended at the mobile station from the diffraction point (in degrees)
- r is the radial distance from the rooftop to the mobile station (in m).

1.1.1.3 Path Loss Model for Outdoor to Indoor and Pedestrian Environment

This environment is served by microcells with radius comprised between 100 m and 1000 m. Base stations with low antenna heights are located outdoors; pedestrian users are located on streets and inside buildings and residences. Log-normal shadow fading with a standard deviation of 10 dB is reasonable for outdoors. The path-loss expression is given as the sum of free space loss, L_{fs}, the diffraction loss from rooftop to the street, L_{rts}, and the reduction due to multiple screen diffraction past rows of buildings, L_{msd}. In this model, L_{fs} and L_{rts} are independent of the base station antenna height, whereas L_{msd} is dependent on whether the base station antenna is below or above building heights. In general the model is given as

$$\overline{L_p(d)} = L_{fs} + L_{rts} + L_{msd}. \tag{1.6}$$

Given a mobile-to-base separation d, the free space loss between them is given by

$$L_{fs} = -10 \log \left(\frac{\lambda}{4 \pi d} \right)^2.$$ (1.7)

The diffraction from the rooftop down to the street level gives the excess loss to the mobile station:

$$L_{rts} = -10 \log \left(\frac{\lambda}{2 \pi^2 r} \left(\frac{1}{\theta} - \frac{1}{2 \pi + \theta} \right)^2 \right).$$ (1.8)

The multiple screen diffraction loss depends on the relative height of the base station antenna as being either below or above the mean building heights. It is given by

$$L_{fs} = -10 \log \left(\frac{d}{R} \right),$$ (1.9)

where R is the average separation between buildings.

1.2 Small-Scale Fading Characterization and Channel Model

Small-scale fading occurs due to the coherent superposition of a great number of multipath components, each having a different phase variation over time or frequency. In built-up areas, fading occurs because the height of the mobile antennas are well below the height of surrounding structures, so there is no single line of sight path to the base station. Even when a line of sight path exists, multipath still occurs due to reflections from the ground and surrounding structures. The frequency response of a multipath channel is frequency selective. The vector sum of each multipath component varies from one frequency to another.

There are two aspects in characterizing the small-scale fading encountered in wireless channels. One focuses on the received signal statistical behavior in terms of envelope and phase. The probability density function of the envelope and the phase of the received signals are reported to cover this aspect. Such characterization is useful for network planning and system capacity estimation.

The other analysis consists of characterizing the channel impulse response, or frequency response, in time and frequency to get an insight on the channel impairments and behavior. The radio channel can be viewed as a doubly dispersive nonlinear time-varying filter. This is additionally useful for receiver design. The channel induces multipath delay spread on the received signal by creating multiple delayed copies of the transmitted signal. The power delay profile is used to characterize this multipath phenomenon. It represents the distribution of energy or power over the duration of the multiple paths. This time interval is known as the delay spread. Multipath induces the frequency selective nature of the radio propagation channel, and the associated delay spread informs on the correlation of signal fading with respect to carrier frequency separation. This in fact leads to the notion of coherence bandwidth when the multipath phenomenon is analyzed in the frequency domain.

In parallel, mobile speed causes Doppler spreading of the carrier frequencies and leads to the notion of coherence time, duration over which the channel can be considered as essentially stationary. Received signals observed at time instants separated by more than the coherence time fade independently since the channel state will have typically changed beyond such interval. This characterization simply reflects user or environment mobility effects.

First an explanation is given of the signal envelope characterization through its probability density function as a function of the environment and propagation conditions.

1.2.1 Statistics of the Received Signal Envelope

In the time domain, the propagation channel can be modeled as a filter with impulse response $h(t; \tau)$. Assuming that $z(t)$ is the complex envelope of the input signal, the filter output $w(t)$ can be expressed as

$$w(t) = \int_{-\infty}^{\infty} h(t; \tau) z(t - \tau) d\tau. \tag{1.10}$$

This impulse response depends on both the observation time t and the delay τ and can be represented as

$$h(t; \tau) = \sum_{l=1}^{L(t)} A(\tau_l(t)) e^{j\phi(\tau_l(t))} \delta(t - \tau_l(t)). \tag{1.11}$$

The gain on a given path l, $A(\tau_l(t))$ and the phase $\phi(\tau_l(t))$ depend on the statistical characteristic of that path. The amplitude and phase distribution of the received signal depend primarily on the presence or absence of a line of sight (LOS) component.

- ■ When the receiver is located in a rich scattering channel and there is no LOS between the transmitter and the receiver, the signal is impinging the receiver antenna from many directions and is the sum of a large number of uncorrelated components. The composite signal can be decomposed in Cartesian coordinates into an in phase and quadrature component, each approximated by Gaussian random variables of equal variance. The resulting received signal envelope follows consequently a Rayleigh distribution:

$$f_A(a) = \frac{a}{\sigma^2} e^{-\frac{a^2}{2\sigma^2}}, \quad a \geq 0. \tag{1.12}$$

Assuming uncorrelated quadrature components, the received signal phase is uniformly distributed in the interval $[0,2\pi]$.

■ In practice, however, occasionally a dominant incoming wave can be a LOS component or a strong specular component. In these situations, the received signal envelope obeys the well-known Rician distribution

$$f_A(a) = \frac{a}{\sigma^2} I_0\left(\frac{a\rho}{\sigma^2}\right) e^{-\frac{a^2+\rho^2}{2\sigma^2}}, \quad a \geq 0, \tag{1.13}$$

where I_0 is the zeroth order modified Bessel function of the first kind. The average power is given by

$$E(A^2) = \rho^2 + 2\sigma^2. \tag{1.14}$$

The parameter $K = \rho^2/2\sigma^2$ gives the ratio between the power in the direct line of sight component to the power of all other delayed paths. When $K = 0$ the channel exhibits a Rayleigh fading behavior. For $K \rightarrow +\infty$, the channel does not exhibit fading at all since the LOS component is dominant. Due to the presence of the LOS component, the phase is no longer uniformly distributed:

$$f_\Phi(\phi) = \frac{1}{2\pi} e^{-\frac{\rho^2}{2\sigma^2}} \left[1 + \sqrt{\frac{\pi}{2}} \frac{\rho \cos \phi}{\sigma} e^{\frac{\rho^2 \cos^2 \phi}{2\sigma^2}} \left(1 + erf\left(\frac{\rho \cos \phi}{\sqrt{2}\sigma}\right)\right)\right],$$

$$|\phi| \leq \pi. \tag{1.15}$$

A purely empirical model that matches more closely experimental data than the Rayleigh or Rician fading models is the Nakagami model. The Nakagami distribution is given by

$$f_A(a) = \frac{2m_m a^{2m-1}}{\Gamma(m)\Omega^m} e^{-\frac{ma^2}{\Omega}}, \quad a \geq 0, m \geq 1/2, \tag{1.16}$$

where $\Omega = E(A^2)$ and $\Gamma(.)$ is the gamma function.

For $m = 1/2$, the Nakagami distribution gives the Rayleigh distribution. For $m = (K + 1)^2/(2K + 1)$, it provides a good approximation of the Rician distribution.

When $m \rightarrow +\infty$, the Nakagami distribution becomes an impulse and indicates the absence of fading. The Nakagami model is often used in analytical modeling and analysis not only because it reflects experimental data well but also because it embeds both the Rayleigh and Rician models as specific cases.

The radio propagation channel can also be characterized by focusing on the filter response rather than on the received signal envelope to gain additional insight on the impairments caused by the channel. As stated previously, the analysis of the channel response reveals a frequency-selective channel due to multipath propagation and Doppler spreading due to user speed.

1.2.2 Characterization of the Radio Channel Response

To reveal the characteristics of the propagation channel, the channel correlation function and the power spectra need to be derived and related to the multipath and the Doppler effects. In the literature, several contributions have consequently derived a number of useful correlation functions and power spectral densities that define the characteristics of fading multipath channels. For details see, for instance, [3–7].

The channel correlation expressions are derived with the assumptions that the multiple paths are uncorrelated and that the low-pass impulse response of the channel is stationary. This leads to the key assumption of uncorrelated scattering. The starting point of the analysis is the autocorrelation function of the equivalent low-pass impulse response of the filter or channel

$$\frac{1}{2} E\, b^*(\tau_1; t) b(\tau_2; t + \Delta t). \tag{1.17}$$

This autocorrelation function in the case of uncorrelated scattering corresponds to the average power output of the channel as a function of time-delay spread τ. In fact, this is called the multipath intensity profile. The range of values of the delay τ for which the function is nonzero is called the multipath spread of the channel. A typical behavior for the channel autocorrelation function is depicted in Figure 1.4 [3].

An analogous characterization of the time varying multipath channel is found in the frequency domain. By taking the Fourier transform of the equivalent low-pass impulse response, the time-variant transfer function is obtained. To reveal the frequency-selective fading nature of the channel and to understand that it is directly related to the multipath phenomenon,

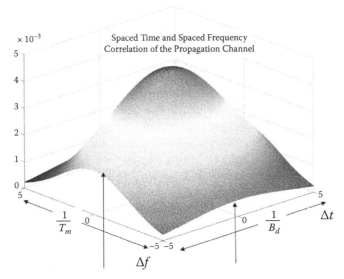

× 10⁻³

Spaced Time and Spaced Frequency
Correlation of the Propagation Channel

- Induced by Multipath and corresponds to Fourier Transform of Multipath Intensity Profile
- Defines Channel Coherence Bandwidth:

$$B_c \approx \frac{1}{T_m}$$

- Induced by Terminal Speed and corresponds to Fourier Transform of Doppler Power Spectrum
- Defines Channel Coherence Time:

$$T_c \approx \frac{1}{B_d}$$

Figure 1.4   Fourier transform of the spaced time and spaced frequency correlation of the propagation channel.

it suffices to derive the autocorrelation function of this Fourier transform at two different frequencies:

$$\frac{1}{2}EH^*(f_1;t)H(f_2;t+\Delta t). \tag{1.18}$$

This is in fact the Fourier transform of the multipath intensity profile. Looking at the frequency spacing $\Delta f = f_2 - f_1$, this autocorrelation function reveals the spaced-frequency and spaced-time correlation behavior of the channel. It can easily be obtained in practice by sending a pair of sinusoids separated by Δf and cross correlating the separate received signals with a relative delay of Δt. Setting $\Delta t = 0$, (1.18) is identified as the Fourier transform of (1.17) with respect to variable τ, the delay spread. As this function is an autocorrelation in the frequency domain, it provides a measure of the frequency coherence of the channel. From the time–frequency duality, the reciprocal of the multipath spread T_m is a measure of the coherence bandwidth of the channel. That is, $B_c \approx \frac{1}{T_m}$ as depicted in Figure 1.4. This indicates that frequencies with spacing larger than B_c fade independently. Signals with bandwidths smaller than the channel coherence bandwidth will experience a frequency nonselective channel. If the signal bandwidth is larger than B_c, the received signal will be severely distorted by the channel.

Other time variations of the channel are measured by parameter Δt. So far only variable τ, the delay spread, has been considered. These time variations in the channel are evidenced as a Doppler spreading and a possible Doppler shift of the carrier frequencies or spectral lines. The Fourier transform of the spaced-frequency and spaced-time correlation in (1.18) taken with respect to variable Δt provides this information. This Fourier transform is depicted in Figure 1.4 as $S_b(\Delta f; \lambda)$. The frequency-domain parameter is λ in this case. By setting $\Delta f = 0$ in this expression, the analysis is conducted only in the Doppler frequency domain. This function therefore reveals a power spectrum that gives the intensity as a function of the Doppler frequency λ. In fact, $S_b(0; \lambda)$ or $S_b(\lambda)$ is known as the Doppler power spectrum of the channel. The range of values where $S_b(\lambda)$ is nonzero corresponds to the Doppler spread of the channel, B_d. Since the Doppler power spectrum is the Fourier transform of the spaced-frequency and spaced-time correlation in Δt, the reciprocal of the Doppler spread B_d is a measure of the coherence time of the channel by duality

$$T_c \approx \frac{1}{B_d}. \tag{1.19}$$

References

1. ITU Recommendation. 1997. Guidelines for Evaluation of Radio Transmission Technologies for IMT-2000. ITU-R M.1225, 2450.
2. "EVOLUTION OF LAND MOBILE RADIO (INCLUDING PERSONAL) COMMUNICATIONS = COST 231." http://www.lx.it.pt/cost231/.
3. Proakis, John G. 2001. *Digital Communications*, 4th ed., New York: McGraw-Hill.
4. Jakes, W. C. ed. 1974. *Microwave Mobile Communications*. New York: John Wiley and Sons.
5. Rappaport, Theodore S. 2002. *Wireless Communications: Principles and Practice*. New York: Prentice Hall.
6. Parsons, J. D. 1985. Radio Wave Propagation. In *Land Mobile Radio Systems*, ed. R. J. Holbeche. London: Peter Peregrinus.
7. Parsons, J. D. 1992. *The Mobile Radio Propagation Channel*. New York: John Wiley and Sons.

Chapter 2

CDMA in Cellular Systems

Code division multiple access (CDMA) is a key radio access technology used in the development of third-generation wireless systems in response to a worldwide growth and demand in mobile communications. This technology, based on spread spectrum communication techniques, inherently fulfills most of the requirements for future generation wireless systems and achieves better bandwidth efficiency. Since CDMA signals occupy larger bandwidth, they are resilient to interference and show smooth capacity degradation as more users are accepted and served. These attractive features of CDMA enable higher bit rates over the air interface, offer better coverage, and quality of service to more users. This chapter describes the CDMA concept and design aspects and highlights the benefits and characteristics of the CDMA technique. The chapter then focuses on CDMA as used in UMTS by emphasizing code construction and describing reception using Rake receiver structures. The chapter serves also as background material for Chapters 3 and 4 addressing the UMTS network and the HSDPA technique used for enhancing UMTS bit rates, radio quality, and capacity.

2.1 CDMA Principle

The multiple-access techniques such as frequency division multiple access (FDMA), time division multiple access (TDMA), or code division multiple access (CDMA) are the keys of the actual cellular systems, 2 and 3G. In fact, cellular systems are conceived as a whole of resources, time–frequency–power, to be shared by a given number of users. FDMA and TDMA systems are considered as a degree of limited freedom since the frequency and the time are multiplexed among users. CDMA systems are based on a completely different approach: All resources are allocated to all simultaneous users, and the transmission power is controlled to maintain a given value of signal to noise ratio (SNR) with the minimum required power. This use of all time–frequency degrees of freedom is possible by using the direct sequence spread spectrum technique.

One of the definitions of a spread spectrum signal is given by Massey in [1]: *"A spread spectrum signal is a signal whose Fourier bandwidth is substantially greater than its Shannon bandwidth"* where the Shannon bandwidth is the bandwidth that the signal needs and the Fourier bandwidth is the bandwidth that the signal uses. In other words, a spread spectrum signal is a signal that uses much more bandwidth than it needs. Figure 2.1 shows a narrowband signal and its bandwidth after spreading.

Since all users occupy the same bandwidth and transmit signals at the same time, a signal of each user should have a particular characteristic to be separated from those of other users. This is often achieved by direct sequence codes that are orthogonal to enable separation of signals at the

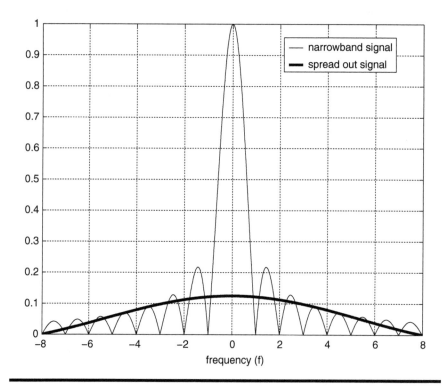

Figure 2.1 Bandwidth of the signal before and after spreading.

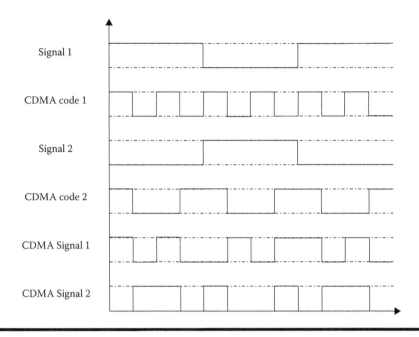

Figure 2.2 Spreading of signals using two orthogonal codes.

receiver. The codes also perform spreading of the signals to provide resistance to interference and robustness to channel impairments. The CDMA codes used in wireless systems are explained in detail in Section 1.3.

Figures 2.2 and 2.3 present an example of two signals spread out by two orthogonal codes. The despreading to restore original signals at the receiver is accomplished by correlating the spread signals by the associated spreading code. Since codes are orthogonal, original signals are separated at the receiver.

2.2 Benefits of CDMA

The main benefits of using a CDMA-based system can be summarized as follows [2]:

■ The CDMA signal experiences a wideband fading channel due to its bandwidth spreading. This can reduce the fade margin needed to achieve reliable radio coverage and increase the signal-to-noise ratio (SNR) of the original signal at the receiver.

■ A CDMA-based system is not a degree of a freedom-limited system [2]. All users can benefit from the full bandwidth at the same time. A higher bit rate per user can then be achieved.

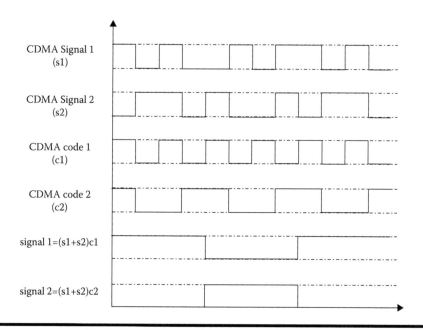

Figure 2.3   Despreading of received signals.

■ By transmitting the signal over the entire bandwidth and by using a pseudonoise (PN) CDMA code, the original signals of each user can be seen as pseudowhite noise. By averaging the interference of all users—since all users exploit the same bandwidth—the interference fluctuations can be limited, which increases the link reliability and makes use of the full bandwidth more efficient.

■ Since the same bandwidth is used in all the network, a user at the cell border can maintain multiple connections in parallel with several base stations. This can increase the cell capacity and the connection quality in case of handover. The name *soft handoff* or *macrodiversity*, explained in the next chapter, is given to this type of handover.

■ CDMA-based systems provide soft capacity, which is not the case in nonCDMA-based systems (e.g., Global System for Mobile, or GSM). In TDMA–FDMA systems, the cell capacity is limited by the number of frequency–time slots pairs. Such systems can only provide hard capacity. In a CDMA system the cell capacity depends only on the signal-to-interference-plus-noise ratio (SINR) of each user and consequently on the overall interference level in the system. This makes the cell capacity soft, and the number of users in the system depends on their positions, quality of service (QoS) requirements, and radio channel conditions.

To benefit from the CDMA characteristics of soft capacity and link relia-bility, adequate interference management must be achieved. This can be performed at the link level by some enhanced receiver structures called multiuser detection (MUD) used to minimize the level of interference at the receiver by removing interference of other users as much as possi-ble before recovering the appropriate signal. At the network level, a good management of the interference can be provided by an enhanced power control and associated call admission control (CAC) mechanisms.

2.3 CDMA Codes

To separate users in a CDMA-based system, orthogonal codes must be used. To make the interference seen by any user as similar to white Gaussian noise as possible, CDMA codes should be selected as long pseudonoise sequences. These two characteristics are typically achieved by using a code constructed by the concatenation of two codes: orthogonal code (Wash code) and scrambling code (pseudorandom code).

2.3.1 Orthogonal Codes

The orthogonal codes used in CDMA systems are based on Wash functions. Wash functions, invented by Wash in [3], are a set of orthogonal functions taking a value of $+1$ and -1, except at a finite number of discontinuity points called jumps where it takes the value zero. The Wash functions have an order, N, which is the number of orthogonal functions. Wash function of order N, denoted by $f_j(t)$ where $t \in [0; T]$ and $j = 0, 1, \ldots, N-1$, can be defined as follows [3,4]:

- $f_j(t) = 1$ or -1 except at the discontinuity points.
- $f_j(0) = 1$ for all j.
- $f_j(t)$ has precisely j sign changes in the interval $[0; T]$.
- $\int_0^T f_j(t) f_k(t) dt = T\delta_{jk}$, where δ_{jk} is the kronecker function.
- Each $f_j(t)$ is either odd or even.

The order N of the Wash functions and its relation to the spreading size have a crucial influence on the design of any CDMA network (e.g., number of users, transmission power, bit rate per user). Therefore, the generation method of the Wash functions has a key role in the conception of any CDMA system. The Wash functions are typically constructed using [4–11]

- The Rademacher functions [12]
- The Hadamard matrices [13,14]

■ The exploitation of the symmetry properties of Wash functions themselves

In this section, particular attention is given to Hadamard matrices used in the third-generation wireless systems, especially the Universal Mobile for Telecommunications System, to generate the Wash functions.

The Hadamard matrices [13,14] are a set of square arrays of $+1$, -1 whose rows and columns are mutually orthogonal. Each element of these matrices is called a *chip*. These matrices, denoted by H_N, can have sizes of $N = 2^k$ where $k = 0, 1, 2, 3, \ldots$. Another characteristic of the Hadamard matrices is $H_{2^p 2^q} = H_{2^p} \times H_{2^q}$ where $p, q = 0, 1, 2, \ldots$. These characteristics simplify the construction of the Hadamard matrices as follows:

$$H_{2N} = \begin{bmatrix} H_N & H_N \\ H_N & -H_N \end{bmatrix}. \tag{2.1}$$

Using this structure of Hadamard matrices, it is easy to verify that $H_N \times H_N^T = N I_N$, where H^T is the transpose matrix and I_N is the $N \times N$ identity matrix. Note that $H_1 = [1]$.

Consequently, by using the Hadamard matrices to generate the Wash functions, the dimension size of these matrices reflects the spreading order (spreading factor, or SF) and the Wash functions order as follows. The matrix rows constitute the Wash functions codes used to spread the narrowband signals. The order of these Wash functions (i.e., the number of orthogonal codes) is then equal to the Hadamard matrix column size, i.e., $N = 2^k$ where $k = 0, 1, 2, 3, \ldots$. Also, the narrowband signal is spread out by correlating each symbol by a given Wash function (i.e., by the chips of a given Hadamard matrix row). The so-called *spreading factor*, which is the order of the bandwidth spreading, is then given by $\text{SF} = \frac{\text{spread-out bandwidth}}{\text{original bandwidth}} = \frac{\text{symbol duration}}{\text{chip duration}}$. This expression of SF explains the fact that the spreading factor is equal to the Hadamard matrix row length, i.e., $N = 2^k$. Since the Hadamard matrices are square, the Wash function order and the spreading factor are equal, which explains the fact that in UMTS, SF indicates not only the bandwidth spreading order but also the number of maximal available orthogonal codes. Note that an increase of the SF value results in a decrease of the user bit rate since the overall carrier bandwidth is limited. In UMTS, SF can take values in powers of two varying from 1 to 512. To support variable data rates, the CDMA air interface can allow selectable spreading factor (i.e., selectable Hadamard matrix dimension), using the flexible structure of Wash–Hadamard functions. This family of variable Wash–Hadamard matrices dimension, used in UMTS, is called orthogonal variable spreading factor (OVSF) codes. An example of OVSF codes is presented in Figure 2.4 [15].

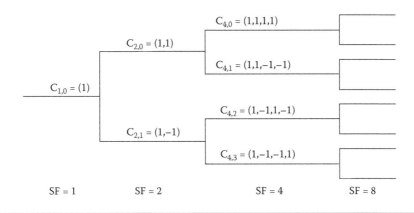

Figure 2.4   Example of OVSF codes tree.

2.3.2   Scrambling Code

Scrambling is applied on top of spreading and does not change rate of the spread-out signal. This code provides improvement in signal auto-and cross-correlation properties. In fact, the Wash–Hadamard codes are very sensitive to the synchronization between codes and lose orthogonality if they are not time aligned. This can occur in the uplink between users and in the downlink between base stations, especially in UMTS, since the base stations are not synchronized in the downlink and the users are not synchronized in the uplink. The improvement of the autocorrelation is required for initial synchronization and for reliable separation between the multipath channel components [16]. Note that a good autocorrelation indicates a good randomness of the signal. Therefore, the improvement of the autocorrelation can be performed by making the spreadout signal as random looking as possible, via modulating it onto a pseudonoise sequence. Since autocorrelation and cross-correlation cannot be improved simultaneously, the role of PN sequences (scrambling codes) is to achieve a good autocorrelation of the spreadout signal with the smallest possible cross-correlation. The presence of some cross-correlations explains the existence of interference between channels in CDMA systems: The interference depends on the cross-correlation.

In general, PN sequences are generated with a linear feedback shift register generator. The output of the shift register cells are connected through a linear function formed by exclusive-OR (XOR) logic gates into the input of the shift register [16].

PN sequences can be either short or long. The long sequence introduces a good randomness in the signal and randomizes as much as possible the interference. The reason to use short sequences is to minimize the complexity of MUD [16]. In UMTS, short scrambling codes are used in the uplink when a MUD is implemented at the base station.

In the literature, a number of PN sequences are presented [4,16,17]: M-sequences used in IS95; Gold sequences; Gold-like sequences, which have a smaller cross-correlation than Gold sequences; Small (S) Kasami sequences; Large (L) Kasami sequences; Very Large (VL) Kasami sequences; and 4-phase Set A sequences, or complex sequences.

The remainder of this section presents a brief description of the scrambling codes used in the UMTS uplink and downlink channels.

2.3.2.1 Scrambling Codes of UMTS Uplink Channels

In UMTS, all uplink physical channels are scrambled with a complex-valued scrambling code. Two scrambling-code families are used: long and short. Note that there are 2^{24} long and 2^{24} short uplink scrambling codes [15].

According to the 3GPP specification [15], the long scrambling sequences can be generated as in Figure 2.5. In this figure, the resulting sequences, C_1 and C_2, which constitute segments of a set of Gold sequences, are generated from position-wise modulo 2 sum of two binary M-sequences. These two M-sequences are constructed using, the following 2 primitive polynomials respectively: $X^{25} + X^3 + 1$ and $X^{25} + X^3 + X^2 + X + 1$.

Concerning the short scrambling codes, they are generated as depicted in Figure 2.6 where [15]:

- $zn(i)$ is a sequence of length 255 generated according to the relation: $zn(i) = a(i) + 2b(i) + 2d(i)$ modulo 4, $i = 0, 1, \ldots, 254$ [Note that $zn(i)$ is extended to length 256 chips by setting $zn(255) = zn(0)$];
- $a(i)$ is a quaternary sequence generated recursively by the polynomial $g0(x) = x^8 + x^5 + 3x^3 + x^2 + 2x + 1$;

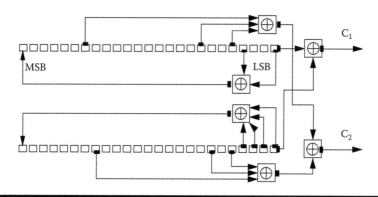

Figure 2.5   Generator of uplink long scrambling sequences. [From 3GPP TS 25.213 V6.4.0. 2005, Spreading and Modulation (FDD), Rel. 6.]

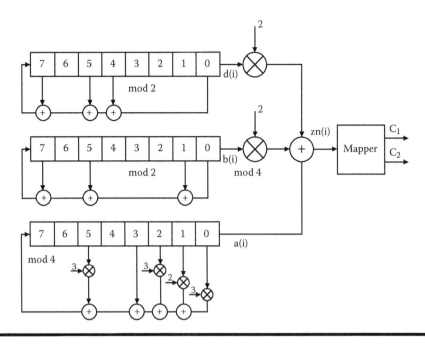

Figure 2.6   Generator of uplink long scrambling sequences. [From 3GPP TS 25.213 V6.4.0. 2005, Spreading and Modulation (FDD), Rel. 6.]

- $b(i)$ is a binary sequence generated recursively by the polynomial $g1(x) = x^8 + x^7 + x^5 + x + 1$;
- and $d(i)$ is a binary sequence generated recursively by the polynomial $g2(x) = x^8 + x^7 + x^5 + x^4 + 1$.

2.3.2.2   Scrambling Codes of UMTS Downlink Channels

In the downlink, the scrambling code sequences are generated by combining two real sequences into a complex sequence. The real sequences, which constitute segments of a set of Gold sequences, are generated from position-wise modulo 2 sum of two binary M-sequences. These two M-sequences are constructed using, the following 2 polynomials respectively: $X^{18} + X^7 + 1$ and $X^{18} + X^{10} + X^7 + X^5 + 1$. The generator of the scrambling-code sequences is depicted in Figure 2.6 [15].

Consequently, a total of $2^{18} - 1$ scrambling codes can be generated. However, not all the scrambling codes are used. The scrambling codes are divided into 512 sets each of a primary scrambling code and 15 secondary scrambling codes. Further details on the generation of the scrambling codes can be found in Reference [15].

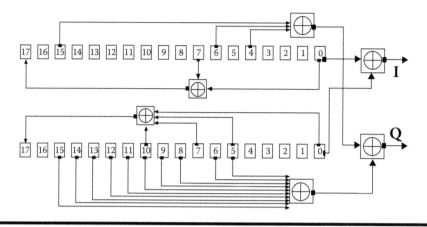

Figure 2.7 Generator of downlink scrambling sequences. [From 3GPP TS 25.213 V6.4.0. 2005, Spreading and Modulation (FDD), Rel. 6.]

2.4 CDMA Receiver

Many types of receivers can be used to adequately decode the CDMA signals. This section provides a brief description of the classic receiver often used in cellular networks.

The receiver is essentially composed of two entities: inner and outer. The role of the inner receiver entity is to demodulate the physical channels, to demultiplex the data symbols, and to pass them to the outer entity wherein several processing functions will be applied on these signals such as deinterleaving, rate matching, channel decoding, tail bits discard, and CRC check. The classic inner receiver contains several entities such as the Rake module, the synchronization–acquisition module, the synchronization–tracking module, and the channel–estimation module. In this section, particular attention is given to the Rake receiver since it is the key of the inner receiver as well as the most popular receiver module described in the literature.

The Rake receiver, proposed by Price and Green in [18], consists of a receiver capable of tracking the multipath channel structure and of increasing the diversity order at the reception [4]. Thanks to scrambling codes, the CDMA signal has a pseudorandom character, with an autocorrelation whose width is on the order of $1/W$ where W is the bandwidth of the spread signal. By using an appropriate receiver, this signal characteristic allows the isolation of multipath signal components, separable by more than $1/W$, for coherent combining [16]. Multipath components with delays greater than the chip time T_{cb} (approximately equal to the inverse of the spreading bandwidth W) apart can be resolved by the receiver. The number of resolvable path is $N_r = T_d/T_{cb}$, where T_d is the maximum excess delay from the first arriving path. To limit the intersymbol interference (ISI), the spreading code must operate at a higher processing rate, or chip rate, to have $T_d \ll T_c$. This simplifies the equalizing and the detection structure.

Consequently, the Rake receiver consists of a bank of cross-correlators, called fingers, each with successive delays of $1/W$. Each finger extracts from the total received signal the portion corresponding to the multipath component arriving at a particular delay. The weighted outputs of the Rake fingers are then combined using maximum ratio combining (MRC). The number of Rake fingers depends on the channel profile and the chip rate. More Rake fingers are needed to catch all the energy from the channel; however, a very large number of Rake fingers leads to combining losses. The number of fingers in the CDMA system varies between two and five.

An example of the Rake receiver architecture is depicted in Figure 2.8, where four fingers are used. In this figure, input signals are transmitted to the four fingers and are preprocessed by an input stage. Then, each finger descrambles and despreads its respective signal flows. To extract the data signal from a particular path, a matched filter is implemented in each finger. An interface block is used to provide a resynchronization process, which is needed to allow a time-coherent combination. Finally, a combiner module is used to combine the weighted outputs of each finger. Note that the RSched is a control block specific to one Rake receiver. It manages the exchanges between the other external parts of the Rake receiver and the logic-blocks Rake receiver. Another task of the RSched is to control the internal control signals of the Rake receiver.

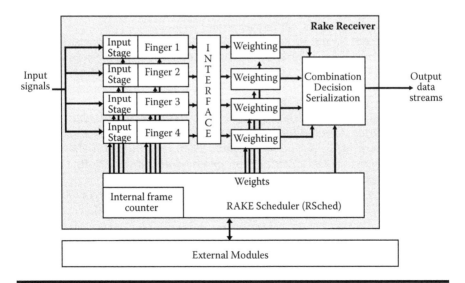

Figure 2.8 Simplified architecture of the Rake receiver.

References

1. Massey, J. L. 1995. Information Theory and Spread Spectrum. In *Code Division Multiple Access Communications*, ed. S. G. Glisic and P. A. Leppanen. Kluwer Academic Publishers, Dordrecht, The Netherlands, 32.

2. Tse, David, and Pramod Wiswanath. 2005. *Fundamentals of Wireless Communication*, Cambridge, UK: Cambridge University Press.

3. Wash, J. L. 1923. A Closed Set of Normal Orthogonal Functions. *American J. Mathematics* 45:5–24.

4. Lee, J. S., and L. E. Miller. 1998. *CDMA Systems Engineering Handbook*. Artech House, Boston.

5. Corrington, M. S. and R. N. Adams. 1962. Advanced Analytical and Signal Processing Techniques: Applications of Wash Functions to Non-linear Analysis. DTIc acc. no. AD-277942, April.

6. Schreiber, H. H. 1970. Bandwidth Requirements for Wash Functions. *IEEE Trans. on Information Theory*. 16:491–2 (July).

7. Harmuth, H. 1968. A Generalized Concept of Frequency and Some Applications. *IEEE Trans. on Information Theory*. 14:375–82 (May).

8. Harmuth, H. 1969. *Transmission of Information by Orthogonal Functions*. New York: Springer-Verlag.

9. Swick, D. A. 1969. Wash Functions Generation. *IEEE Trans. on Information Theory*. 15:167 (January).

10. Proceedings of the 1970 Symposium on Applications of Wash Functions, Naval Research Laboratory, Washington, DC, March–April.

11. Proceedings of the 1973 Symposium on Applications of Wash Functions, The Catholic University of America, Washington, DC, April 16–18.

12. Alexits, G. 1961. *Convergence Problems of Orthogonal Series*. New York: Pergamon Press.

13. Golomb, S. W. 1964. *Digital Communications with Space Applications*. Englewood Cliffs, NJ: Prentice-Hall.

14. Ahmed, N. and K. Rao. 1972. Wash Functions and Hadamard Transform. Proceedings of the Symposium on Applications of Wash Functions, The Catholic University of America, Washington, DC, March 8–13.

15. 3GPP TS 25.213 V6.4.0. 2005. Spreading and Modulation (FDD), Release 6, September.

16. Ojanpera, T. and R. Prasad. 2001. *WCDMA: Towards IP Mobility and Mobile Internet*. Artech House Universal Personal Communications series, Boston.

17. Hammons, R. A. and V. J. Kumar. 1993. On a Recent 4-Phase Sequence Design for CDMA. *IEICE Trans. on Communications* E76-B, no. 8: 804–13 (August).

18. Price, R. and P. E. Green, Jr. 1958. A Communication Technique for Multipath Channels. *Proceedings of the IRE* 46:555–70 (March).

Chapter 3

Universal Mobile for Telecommunications System

To meet the requirements of the fast-growing demand for wireless services due to the integration of Internet, multimedia, and mobile communications, third-generation (3G) wireless networks are being developed under the International Telecommunication Union (ITU) initiative by the 3GPP and 3GPP2 (Third Generation Pattern Project). The 3GPP is a joint venture of several international standardization organizations from Europe (ETSI), Japan (ARIB/TTC), the United States (T1P1), South Korea (TTA), and China (China Wireless Telecommunication Standard Group). The 3GPP2 was created in parallel with the 3GGP project, with participation from ARIB, TTC, and CWTS. 3GPP2 focuses more on the IS-95 evolution. Third-generation systems developed by 3GPP and 3GPP2 meet the IMT-2000 (International Mobile Telecommunication 2000) requirements including higher data rate and spectrum efficiency than 2G systems, support of both packet-switched (PS) and circuit-switched (CS) data transmission, wide range of services and applications, quality of service differentiation, and flexible physical layer with variable bit rate capabilities to ease the introduction of new services. The most important IMT-2000 proposals are the Universal Mobile for Telecommunications System (UMTS) and the CDMA2000 as successors respectively to GSM and IS-95 systems.

The UMTS meets the IMT-2000 requirements by supporting a wide range of symmetric/asymmetric services. Users can access traditional public switched telephone networks (PSTN)/integrated services digital network

(ISDN) services as well as emerging IP data communications services including Internet and multimedia applications with unprecedented efficiency and flexibility in wireless communications systems. The wireless technique adopted for UMTS is the code division multiple access (CDMA).

The UMTS air interface, UMTS terrestrial radio access (UTRA), as introduced in 1998 by ETSI, supports two modes: (1) UTRA FDD, based on wideband CDMA (WCDMA) for frequency division duplex (FDD) operation; and (2) UTRA TDD, based on Time Division CDMA (TD/CDMA) for time division duplex (TDD) operation. The UTRA FDD mode uses direct sequence code division multiple access (DSCDMA) technology, with a chip rate of 3.84 Mchips/s, to operate in paired spectrum bands (5 MHz each for uplink and downlink). The use of WCDMA allows the increases multipath diversity and results in signal quality and user bit rate improvements. The bandwidth used for uplink and downlink are, respectively, [1920, 1980] and [2110, 2170] MHz. The TDD mode uses TDMA technology and time-slot spreading for operation in the unpaired bands: 5 MHz shared dynamically between uplink and downlink. Mode selection depends on spectrum availability and type of coverage: symmetric or asymmetric. The bandwidth reserved for this mode are [1900, 1920] and [2010, 2025].

The evolution of the 3G system inside the 3GPP project has been organized and scheduled in phases and releases (99, 4, 5, and 6). The first release was the UMTS Release 99, introduced at the end of 1999. It supports high-speed services by providing data rates up to 2 Mbps depending on speed and coverage. This release defines and specifies seamless transitions from existing GSM networks to enable transparent intersystem (GSM/WCDMA and vice versa) handovers. This chapter focuses on Release 99 of the WCDMA FDD mode of UMTS Release 99 as deployment has already started in Europe and Asia. Future releases are presented when HSDPA is addressed in ensuing chapters.

In Release 4, completed in March 2001, improvements over Release 99 were added to the 3G standards including the introduction in the TDD mode of an additional chip rate option of 1.28 Mchips/s to be used in addition to the initial 3.84 Mchips/s rate specified in Release 99. Enhancements in this release allow also the packet data convergence protocol (PDCP) layer (see Section 3.8) to support new IP header compression algorithms in particular the protocol described in RFC 3095 [1]. On the core network side, the main improvement is the separation in the mobile switching center (MSC) of the user and control planes respectively into the media gateway (MGW) and MSC.

In Release 5, the main improvement is the development of the high-speed downlink packet access (HSDPA) to achieve higher aggregate bit rates on the downlink. HSDPA is based on the introduction of new enhancement techniques such as fast scheduling, adaptive modulation and coding, and hybrid automatic repeat request (HARQ). HSDPA relies on a distributed architecture where more intelligence is introduced in the node B (base station) to handle packet data processing and thereby allowing faster scheduling and retransmissions mechanisms. HSDPA is described in the next chapter.

In addition to HSDPA, release 5 introduces the IP multimedia subsystem (IMS) to support IP based transport and service creation.

In order to support multiple services at higher bit rates, Release 6 and beyond are focusing on the introduction of new features and enhancements including [2,3]:

- New transport channel in the uplink called enhanced dedicated channel (E-DCH) to improve coverage and capacity in order to provide higher bit rate services
- Advanced antennas technologies, such as beamforming and multiple input multiple output (MIMO)
- Introduction of new higher bit rate broadcast and multicast services (see Section 3.9)
- Addition of new features to the IMS
- Possibility of new frequency variants use (for WCDMA), such as use of the 2.5 and 1.7/2.1 GHz spectrum
- Improvements of GSM/EDGE Radio Access Network (GERAN) radio flexibility to support new services

This chapter focuses essentially on the WCDMA FDD mode of the UMTS Release 99 as the main third-generation technique whose deployment has started in Europe and Asia. Features and enhancements of the subsequent releases are also indicated.

The UMTS services and their quality of service (QoS) requirements are described in Section 3.1. The general architecture of a UMTS network is introduced in Section 3.2. The rest of this chapter focuses essentially on the universal terrestrial radio access network (UTRAN) entities and protocols. Section 3.3 presents the radio interface protocol architecture used to handle data and signaling transport between the user, UTRAN, and the core network. Sections 3.4 to 3.10 describe these protocol layers. Section 3.11 provides description of the automatic repeat request (ARQ) protocol used in the UMTS at the radio link control (RLC) layer. This protocol has a major impact on the performance of transport control protocol (TCP) in the UMTS network (as explained later in Chapters 5–7). Power control and handover for the UMTS air interface are respectively described in Sections 3.12 and 3.13. Finally, Section 3.14 presents a basic mathematical analysis of the UMTS capacity and system performance.

3.1   UMTS Services

UMTS is required to support a wide range of applications with different QoS requirements (bit rate, errors rate, delay, jitter, etc.). These applications can be classified into person-to-person services, content-to-person services, and business connectivity [3]. Person-to-person services consist of peer-to-peer services between two or more subscribers. Depending on the QoS requirements and the type of application, these services can be offered either in the circuit-switched domain (e.g., speech, video telephony) or the packet-switched domain (e.g., Multimedia Messages [MMS], push-to-talk, voice-over IP [VoIP], multiplayer games) of the UMTS network. Content-to-person services consist of content downloading, access to information, as well as broadcast/multicast at high rate. These services include wireless access protocol (WAP), browsing, audio and video streaming, and multimedia broadcast/multicast services (MBMS). Note that MBMS consists of sending audio, streaming or file downloading services to all users or a group of users using specific UMTS radio channels (see Section 3.9). Business connectivity consists of access to Internet or intranet using laptops via the UMTS radio interface. The effect of severe wireless conditions represents a key aspect to address and to assess overall performance. In order to guarantee QoS requirements of UMTS applications and services (using a remote host), several studies have analyzed interactions between layers and explored solutions to mitigate their negative effects. This book focuses on the interaction between 3G (UMTS and HSDPA) protocol layers and data services using TCP.

In the 3GPP, services are classified into groups according to their QoS requirements, which define priorities between services and enables allocation of appropriate radio resources to each service. When cells are heavily loaded, the network blocks or does not accept (in case when new service arrives) services with low priority or services requiring more resources than available. The UMTS system can also delay the transmission of data services that have low-delay sensitivity and allocate the resources to the services that should be transferred quickly (e.g., real-time services).

Four distinct QoS classes are defined in the 3GPP specifications [4]: conversational, streaming, interactive, and background. The conversational class has the most stringent QoS requirements, whereas the background class has the most flexible QoS requirements in terms of delay and throughput.

3.1.1   Conversational Class Applications

Conversational applications refer essentially to multimedia calls such as voice and video telephony. Applications and services such as NetMeeting, Intel VideoPhone, and multiplayer games are good examples of applications that map onto this QoS class [5].

Conversational applications are the most delay-sensitive applications since they carry real-time traffic flows. An insufficiently low transfer delay may result in service QoS degradations. The QoS requirements of this class depend essentially on the subjective human perception of received applications traffic (audio and video), as well as the performance of the used codec (i.e., audio/video source coding). Studies on the human perception of the audio/video traffic have shown that an end-to-end delay of 150 ms is acceptable for most user applications [6]. In the 3GGP specifications [7], it is stated that 400 ms is an acceptable delay limit in some cases but it is preferable to use 150 ms as an end-to-end delay limit. Concerning the time relation (variation) between information entities of the conversational stream (i.e., jitter), the audio/video codec does not tolerate any jitter. According to [7], the jitter should be less than 1 ms for audio traffic. Note that audio traffic is bursty with silent periods between burst depending on the codec, user behavior, and application nature (radio planning). Video traffic is transmitted regularly with variable packet lengths and this results in burstiness in particular when variable bit rate codecs are used. The bit error rate (BER) target (or limit) should vary between $5*10^{-2}$ and 10^{-6} depending on the application [4]. This corresponds to frame error rates (FER) less than 3 percent for voice and 1 percent for video as advocated in [7].

These stringent delay requirements prevent the data link layer from using retransmission protocols. ARQ consists of retransmitting erroneous packets but this results in additional delays that are unacceptable for conversational applications. Conversational traffic is consequently carried over the user datagram protocol (UDP) instead of the TCP. The reliability of TCP is achieved via retransmissions and congestion control that consists of retransmitting packets (subject to errors or congestion). A dynamic transmission window is also used to regulate TCP traffic. These mechanisms result in additional delays. TCP is mostly used for nonreal-time services while UDP that does not use any flow congestion control is used for real-time and conversational traffic classes.

3.1.2 Streaming Class Applications

Streaming applications have less delay requirements than conversational services. The fundamental characteristic of this applications class is to maintain traffic jitter under a specific threshold. Jitter relates to the time relation between received packets. This threshold depends on the application, the bit rate, and the buffering capabilities at the receiver. The use of a buffer at the receiver smoothes traffic jitter and reduces the delay sensitivity of the application. Video sequences are managed by the client at the receiver that plays the sequences back at a constant rate. A typical buffer capacity at the receiver is 5 s. This means that the application streams could not be delayed in the network more than 5 s. In the 3GPP specifications [7], it is indicated that the start-up delay of streaming applications should be less than 10 sec and the jitter less than 2 sec. If the buffer is empty, due to jitter and delay latency, the application will pause until enough packets to play are stored in the buffer. Typical examples of the streaming application softwares are RealPlayer and Windows Media Player that are able to play audio and video streaming.

The most suitable protocol stack to handle streaming services is RTP (real time protocol)/RTCP (real time control protocol)/UDP since it can achieve low delays and low jitter. However, streaming is carried in certain cases over TCP protocols (e.g., network containing firewalls requiring use of TCP, nonlive streaming applications that are completely downloaded before being played).

The use of retransmission mechanisms is acceptable as long as the number of retransmissions and the overall delay are limited. The use of HARQ mechanism in the medium access control-high speed (MAC-hs) sublayer in HSDPA is acceptable. The ARQ protocol of the RLC sublayer, however, is not suitable for this class of applications due to the important delays introduced in the received information. The MAC-hs and RLC sublayers, as well as the ARQ and HARQ mechanisms, will be described later in this chapter and in the next chapter. Concerning the tolerated error rates at the receiver, the target or limit BER can vary between $5 * 10^{-2}$ and 10^{-6} depending on the application as indicated in the 3GPP specifications [4].

3.1.3   Interactive Class Applications

This applications class has lower delay requirements than conversational and streaming classes. It consists of applications with request response patterns such as those of the Web and WAP browsing. In this class, applications are essentially of the server-to-person type where the user requests information and waits for a response from the server in a reasonably short delay. The quantity of transfer information on the downlink is more important than on the uplink in these cases. Since more delay latency can be tolerated, retransmission mechanisms such as ARQ protocol at MAC-hs or RLC sublayers can be used to achieve reliability (error free reception) at the link layer with less radio resource consumption (more details on the use of ARQ in wireless system can be found in Sections 3.11 and 4.6).

Web browsing applications are basically conveyed over TCP, whereas WAP services are carried over UDP or a WAP specific protocol called wireless data protocol (WDP). Web browsing services and TCP protocol are described in Chapter 5, where an example of Web data transfer over TCP is presented. An overview of WAP technologies is also presented in Chapter 6.

Finally, interactive services can tolerate large delays and jitter and high reliability, low-residual error rate, is easily achieved. The residual error rate afforded via ARQ should be less than $6 * 10^{-8}$ and the delay latency less than 4 sec per page [6,7].

3.1.4 Background Class Applications

This class presents the most delay latency tolerance since the destination does not expect the data within a certain time. This application data can be sent in the background of other application classes. Typical examples of this class are e-mail, file transfer protocol (FTP), short messages (SMS), and multimedia messages (MMS). The link carrying these services should present high reliability since these applications need to be received correctly (i.e., payload content must be preserved). The residual bit error rate should be less than $6 * 10^{-8}$ [4]. Therefore, ARQ mechanisms at the MAC-hs and RLC sublayers can be used, as well as reliable transport protocol such as TCP.

An overview of QoS requirements of conversational, streaming, interactive, and background services is presented in Table 3.1 [6,7].

This book focuses on the interaction of applications carried over TCP in UMTS-HSDPA cellular systems. Challenges and analysis described and derived in Chapters 5, 6, and 7 concern only interactive and background classes.

3.1.5 Quality of Service (QoS) Parameters

In addition to the traffic classes, other QoS parameters have been defined in [5,8] to differentiate between services and to achieve the appropriate service quality of each application.

Table 3.1   Overview of Services Classes QoS Requirements [6,7]

Service class	Conversational	Streaming	Interactive	Background
One-way delay	< 150 ms (preferred) and <400ms (limit)	< 5 sec	< 4 sec/page	No limit
Bit error rate	between $5 * 10^{-2}$ and 10^{-6}	between $5 * 10^{-2}$ and 10^{-6}	between $4 * 10^{-3}$ and $6 * 10^{-8}$	between $4 * 10^{-3}$ and $6 * 10^{-8}$
Delay variation	< 1 ms	< 2 sec	N.A	No limit
Use of retransmission mechanism	not used	MAC-hs	MAC-hs, RLC	MAC-hs, RLC
Transport Layer	UDP	UDP and sometimes TCP	TCP (WDP for WAP)	TCP

- *Traffic class (conversational, streaming, interactive, background).*
- *Maximum bit rate.*
- *Guaranteed bit rate.*
- *Maximum service data unit (SDU) size* (at the radio link control layer, see Section 3.7 for more details). This parameter is used in admission control and policing.
- *SDU error ratio* indicates the fraction of erroneous SDUs.
- *SDU format information* indicates all possible sizes of SDUs. This information is used essentially in RLC transparent mode (definition of RLC transparent mode is given in Section 3.7).
- *Delivery order* indicates whether the SDUs are delivered in sequence or not to the upper layers. This parameter has an important impact on the TCP performance in UMTS as we will explain later in this book (Chapters 5, 6, and 7).
- *Residual bit error ratio* indicates the undetected BER in the delivered SDUs to the upper layers (i.e., beyond RLC layer).
- *Delivery of erroneous SDUs* which indicates whether erroneous SDUs at the RLC layer are delivered to the upper layer or not depending on the RLC mode used (see Section 3.7).
- *Discard timer* indicates the time after which an erroneous SDU is not retransmitted (see Section 3.7).
- *Transfer delay* defined in [5,8] as *"the maximum delay for the 95th percentile of the distribution of delay of all delivered SDUs during the life time of the connection."*
- *Allocation/retention priority* indicates the priority or the relative importance of a UMTS connection compared to other connections. This information is used during admission control and resource allocation procedures.
- *Source traffic characteristics.*

3.2   General Architecture

The general architecture of a UMTS network is shown in Figure 3.1 [9–11]. Network elements in this architecture can be grouped into three domains: user equipment domain, UTRAN domain, and core network domain. Domains and entities in each domain are separated by reference points serving as interfaces [3].

This architecture has been conceived to vehicle and manage CS and PS traffic. Thus, the external networks can be divided into two groups: (1) CS networks, such as PSTN and ISDN; and (2) PS networks, such as Internet and X25.

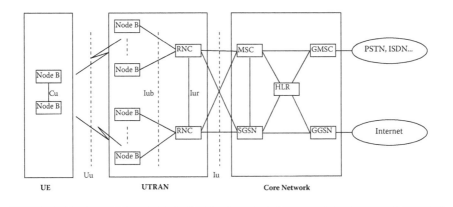

Figure 3.1   General architecture of UMTS network.

3.2.1   User Equipment Domain

User equipment is a device allowing a user access to network services [10]. This domain can be divided into two parts or subdomains called mobile equipment domain and user services identity module domain (USIM), separated by the Cu interface. The user equipment domain is connected to the UTRAN by the Uu interface [9].

The mobile equipment is the terminal that performs radio transmissions and related functions. The USIM is the smart card that contains the user identity and subscription information regardless of the terminal used. This device holds user profile data, procedures, and authentication algorithms that allow secure user identification.

3.2.2   UTRAN Domain

The access network domain provides to the user equipment the radio resources and the mechanisms necessary to access the core network domain. The UTRAN contains entities that control the functions related to mobility and network access. They also allocate or release connections (radio bearers). The UTRAN consists of radio subsystems RNS connected to the core network by the Iu interface. Each RNS includes one RNC and one or more node Bs [9,12].

The node B is in reality the base station. It is the entity that allocates and releases radio channels, partially manages the radio resources, controls the transmission power in the downlink, and converts the data flow between Iu and Uu interfaces. In Release 99, the node B contains procedures that manage only the physical layer such as coding, spreading, transmission and detection, and physical layer signaling [12]. In Releases 5 and 6, intelligence is introduced in the node B, which is then able to perform MAC functions, especially scheduling and HARQ.

The RNC is the main element and the intelligent part of the RNS. The RNC controls the use and the reliability of the radio resources. It performs the functions of the MAC/RLC layer and terminates the radio resource control (RRC) protocol. In the 3GPP [12], three types of RNC have been specified: SRNC (serving RNC), DRNC (drift RNC), and CRNC (controlling RNC).

The SRNC is the entity that holds the RRC connections with the UEs. This entity is the point of connection between the UTRAN and the core network and is involved in the user mobility management within the UTRAN.

When a user equipment moves in the connection state from a cell managed by the SRNC to another associated with a different RNS, the RNC of the new cell is called DRNC. The RRC connection is still handled by the SRNC. In this case, the DRNC serves as a simple relay to forward information between the SRNC and the user equipment.

The CRNC is the RNC performing the control and the configuration of a node B. This entity holds the responsibility of load control in its own cells, as well as the admission control of and code allocation to new users accessing the system.

3.2.3 Core Network Domain

The core network domain is basically inherited from the GPRS network architecture according to a transition phase, from GPRS to UMTS networks, specified in 3GPP. The core network consists of physical network entities integrating both circuit and packet-switched domains. The CN domain provides various support functions for services traffic conveyed over the UMTS system. The services correspond to management of user location formation, control of network features, and transfer mechanisms for signaling [9]. The core network includes switching functions for circuit-switched services via the mobile switching center (MSC) and the gateway MSC. The home location register (HLR) and the visitor location register (VLR) are the databases responsible for handling, respectively, user subscriptions and terminals visiting various locations. To manage packet data services, the packet domain relies on the serving GPRS support node (SGSN) and gateway GPRS support node (GGSN), which serve, respectively, as routers and gateways. The SGSN and the GGSN are involved in the management of session establishment (i.e., packet data protocol contexts) and in the mobility of data services. In certain cases, mobility management is achieved jointly by circuit-switched and packet-switched domain via the cooperation of the SGSN, GGSN, MSC, HLR, and VLR via dedicated interfaces fully described in 3GPP specifications.

This section provides a brief overview of the core network entities. For more details, see [10].

The HLR is a database handling maintenance of user subscription data and profiles. This information is transferred to the adequate VLR or the SGSN in order to achieve location and mobility management. In addition, the HLR provides routing information for mobile calls and short message service (SMS).

The VLR is involved in user location updates in the circuit-switched domain (functions inherited from the GSM architecture). It contains subscriber information required for call and mobility management of subscriber visiting the VLR area. The MSC is a switch mainly used for voice and SMS. It is involved, with the public switched telephone network (PSTN), in the establishment of end-to-end circuit-switched connections via signaling system 7 (SS7). In addition, it is coupled to the VLR to achieve mobility management.

The gateway MSC provides switching for CS services between the core network and external CS networks and is involved in international calls.

The SGSN plays in the packet domain a similar role to the MSC/VLR in the circuit-switched domain. It handles location and mobility management, by updating routing area, and performs security functions and access control over the packet domain.

The GGSN serves as an edge router in the core network to convey data between UMTS network and external packet networks (e.g., Internet). In other words, it has in the packet domain the same role that the GMSC does in the circuit-switched domain. The GGSN is involved in packet data management including session establishment, mobility management, and billing (accounting). In addition, the GGSN includes firewall and filtering of data entering the core network in order to protect the UMTS network from external packet-data networks (e.g., Internet).

3.2.4 Interfaces

The interfaces between the logical network elements of the UMTS architecture are defined in the 3GPP specifications [9,10,13–15]. The main interfaces are Cu, Uu, Iu, Iur, and Iub.

The Cu interface is the electrical interface between the USIM and the mobile equipement. Uu is the WCDMA air interface described later in this chapter.

3.2.4.1 Iu Interface

The interface between the UTRAN and the core network, called Iu and described in [13], is involved in several functions handling control and signaling of mobile calls. These functions include establishment, management, and release of radio access bearers (connections). The Iu interface achieves the connection between the UTRAN and the circuit-switched and packet-switched core network domains. This interface allows the transfer of signaling messages between users and core network and supports cell broadcast service via the Iu broadcast. In addition, the Iu interface supports location services by controlling location reporting in the RNC and by allowing transfer of the location report (including geographical area identifier or global coordinates) to the core network. The Iu interface is also involved in relocation of the serving radio network subsystem (SRNS), as well as intra and inter system handover management. The SRNS relocation is caused in general by a hard handover at the Uu interface that generates a change of radio resources at this interface. Another target for the relocation is to keep all radio links in the same drift radio network controller (DRNC). In this case, radio resources are not altered and data flows remain uninterrupted.

3.2.4.2 Iur Interface

The Iur interface, used between RNCs and described in Reference [14], provides the capability to support essentially four distinct functions: (1) inter RNC mobility, (2) radio resource management, (3) dedicated channel traffic (mobility and management), and (4) common channel traffic (mobility and management). Note that the definition of dedicated and common channels are given in Section 3.4.

The inter RNC mobility function allows the support of radio interface mobility between radio network subscribers (RNSs) including SRNS relocation, packet errors reporting, and inter RNC registration area update. This function does not support any user data traffic exchange between RNCs.

The radio resource management function, introduced in the subsequent Releases 5 and 6 for Iur optimization purposes [3], provides transfer of signaling information between RNCs including node B timing, positioning parameters, and cell information and measurements.

The Iur interface is involved in the mobility management of the dedicated and shared channel traffic. In the case of handover, this interface provides the capability of supporting establishment and release of the dedicated channel in the DRNC of the new cell area. It provides support of the radio links management in the DRNC by allowing the exchange of measurement reports and power control settings. In addition, it achieves transfer of the dedicated channel transport blocks between serving and drift RNCs.

In addition to the dedicated and shared channels, the Iur interface supports the mobility management of the common channel traffic including set-up and release of the transport connection, as well as flow control handling between RNCs. This function has the drawback of introducing more complexity in the Iur interface and generating inefficiency in the resources utilization [3] by splitting the medium access control (MAC) layer into two entities MAC-d and MAC-c located, respectively, in the SRNC and DRNC.

3.2.4.3 Iub Interface

The Iub interface [15], used between the RNC and Node B, supports location handling service and mobility management. It is involved in the admission control of mobile users, resource allocation (e.g., outer-loop power control), connection establishment, and release, as well as handover management. This interface also allows control of the radio equipment and radio frequency allocation in the node B.

Other interfaces are described in the 3GPP specifications [10]. These interfaces include those between MSCs, VLRs, HLR and MSC, HLR and VLR, MSC and GMSC. Interfaces within the packet domain and between the packet and circuit domains are also readily available.

3.3 UTRAN Protocol Architecture

In order to manage and handle mobile calls, data transfer, and signaling information between users and the core network (across the different interfaces described before), the UTRAN comprises nodes that handle the information (data and control) at different protocol layers. Figure 3.2 depicts the

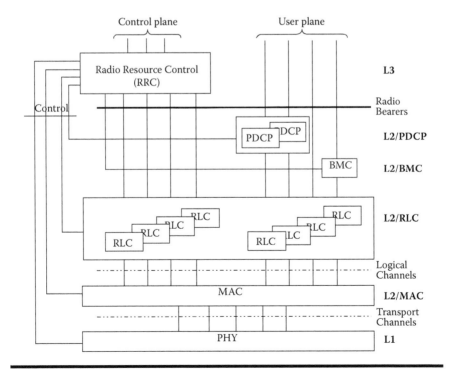

Figure 3.2 Radio interface protocol architecture of UMTS.

UTRAN protocol architecture that consists of several protocol layers visible in UTRAN [16]:

1. Physical layer
2. Data link layer that contains

   ■ Medium access control (MAC)
   ■ Radio link control (RLC)
   ■ Packet data convergence protocol (PDCP)
   ■ Broadcast/multicast control (BMC)

3. Radio resource control (RRC) at the network layer

The UTRAN distinguishes two message planes: (1) user plane to convey user messages and (2) control plane to manage signaling and control messages. In the control plane, the data link layer (layer 2) is split into two sublayers: MAC and RLC. In the user plane, layer 2 contains, in addition to MAC and RLC, two sublayers: PDCP and BMC. Layer 3 (network layer), visible in the UTRAN, contains only one sublayer called the radio resource control (RRC) located in the control plane. Note that call control, session management (SM), and mobility management are transparent to UTRAN and thus not included in the radio interface protocol architecture.

In Sections 3.4 to 3.10 of this chapter, an overview of the UTRAN protocol sublayers is presented. Particular attention is given to MAC and RLC layers that interact with the TCP/IP protocol stack and affect system efficiency and performance.

3.4   UMTS Channels

To carry and manage several traffic types over the air interface, the 3GPP specifications define several channels, each having a specific role in establishing and maintaining sessions in the UMTS access network. These channels can be divided into three groups: logical channels, transport channels, and physical channels.

3.4.1   *Logical Channels*

A logical channel is defined according to the type of information it transports. One can distinguish two classes of logical channels: those for control, and those for traffic.

3.4.1.1   *Logical Control Channels*

Logical control channels are used for the transfer of information in the user plane. The various logical control channels are [16,17]

- Broadcast control channel (BCCH) used on the downlink to broadcast system and network information in all the cells.
- Paging control channel (PCCH) used on the downlink to carry paging information for mobile-terminated calls and sessions.
- Common control channel (CCCH) used on both uplink and downlink to transport signaling information to all users.
- Dedicated control channel (DCCH) used on both uplink and downlink to transport signaling information between the UTRAN. This channel is a spreading code dedicated to a user.
- Shared control channel (SHCCH) used on both uplink and downlink in TDD mode only to transmit control information between UTRAN and mobile stations.
- Multimedia broadcast/multicast service (MBMS) point-to-multipoint control channel (MCCH) used on the downlink to carry MBMS control information from the UTRAN to the UE. (MBMS is described in Section 3.9.)
- MBMS point-to-multipoint scheduling channel (MSCH) used on the downlink to carry MBMS scheduling control information for one or more MTCH channels (MBMS point-to-multipoint Traffic Channel).

Logical traffic channels are used for the transfer of information in the user plane. The three types of specified traffic channels are [16,17]

- Dedicated traffic channel (DTCH) used on both uplink and downlink for the data transmission between the UTRAN and a dedicated user
- Common traffic channel (CTCH) used for the transport of messages to all cell users
- MBMS point-to-multipoint traffic channel (MTCH) used on the downlink to carry MBMS traffic data from the network to the user equipment

3.4.2 Transport Channels

Transport channels are services offered by layer 1 to the higher layers [18,19]. The transport channel is unidirectional (i.e., uplink or downlink) and consists of the characteristics required for the data transfer over the radio interface. For example, the size of the transport block (to transport a data unit of the MAC layer) is one of the transport channel characteristics. Note that the corresponding period to transmit a transport block is known as the transmit time interval (TTI). In Release 99, the TTI can take the values of 10, 20, 40, or 80 ms. For voice services, the TTI is fixed at 10 ms whereas for data services it changes according to the service used. In Release 5, the TTI of the high-speed downlink shared channel (HS-DSCH) has been

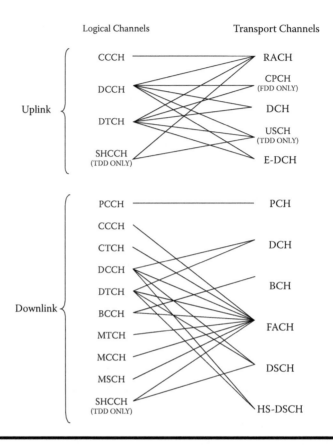

Figure 3.3 Mapping of logical channels onto transport channels.

reduced to 2 ms (for data services) to introduce finer grain control and scheduling in the system. The HS-DSCH channel is described in details in the next chapter.

Transport channels are classified in three groups: dedicated channels, common channels, and shared channels. The mapping of the logical channels onto the transport channels is depicted in Figure 3.3 [16,17].

The dedicated channel is a point-to-point channel used on both uplink and downlink to carry data and control information (from higher layer) between the UTRAN and a specific (dedicated) user. The nature of transmitted information is transparent to the physical layer that carries user data and control information in the same way. However, the physical layer parameters established by the UTRAN depend on the nature of the transmitted information on the channel (i.e., data or control). Note that the dedicated channel can be transmitted over the entire cell or over only a part of the cell. The DCH can support soft handover and fast power control.

In Release 6, enhancements are introduced in the UMTS standard to achieve higher data rate on the uplink. A new transport channel is specified for this purpose to support higher order modulation and HARQ techniques. This channel, called enhanced dedicated channel (E-DCH), is subject to fast power control and node-B controlled scheduling. More details on this channel can be found in [16].

The specified shared channels in UMTS UTRA FDD mode, the specified shared channels are DSCH and HS-DSCH. The downlink shared channel (DSCH) is a downlink transport channel shared by several users. The DSCH is consequently associated with one or several downlink dedicated channels when shared by multiple users. The DSCH is transmitted over the entire cell or over only a part of the cell using, for example, beamforming antennas. The HS-DSCH is a downlink transport channel shared by several user equipments and is associated with one downlink dedicated physical channel (DPCH) and one or several HS-SCCH. This channel introduced to achieve higher data rates over the air interface is presented in detail in the next chapter.

In the TDD mode, the 3GPP specifies [16] another shared transport channel to be used on the uplink. This channel, called uplink shared channel (USCH), is shared by several user equipments carrying dedicated control or traffic data.

A common channel is a unidirectional point to multipoint channel used either on uplink or downlink to transfer information between the UTRAN and one or more users. The main common channels specified in [16] are

- The broadcast channel (BCH) is a downlink transport channel used to broadcast system and cell specific information over the entire cell. This channel has always a single transport format independently of radio environment and type of information.
- The random access channel (RACH) is an uplink transport channel received from users in the entire cell. It contains control information allowing the mobile to access the network via connection or channel requests. The transmission power of this channel is estimated using open-loop power control. Users attempting initial access to network resources via the RACH rely on received downlink channels (especially pilot channels or the BCH) to estimate the required amount of power needed for RACH transmission. The RACH is shared by users using a carrier sense multiple access/collision avoidance (CSMA/CA) technique to manage collisions and contention over the radio interface. In addition to the random access, the RACH could also be used for transmitting very short packets in RLC unacknowledged mode (UM).

- The forward access channel (FACH) is a downlink point-to-point transport channel. The FACH is transmitted over the entire cell and is used to carry signaling information to the user. For example, the FACH is used to carry access grant messages in response to channel requests or random access messages, received from users (that want to access the network) on the RACH.
- The paging channel (PCH) is a downlink transport channel transmitting paging control information toward mobile stations when the system does not know the precise user location. It is typically transmitted in several cells to locate the user. The transmission of the PCH is associated with the transmission of the physical paging indicator channel (PICH) allowing the support efficient sleep-mode procedures. The paging messages awaken terminals from sleep mode.
- The common packet channel (CPCH) is an uplink random access transport channel similar to the RACH. The main difference between these two channels is that the CPCH is used only in connected mode. The CPCH was conceived to transfer larger-size packets whereas the RACH is mostly used for random access or short packets in RLC UM mode. The CPCH is associated with a dedicated channel on the downlink that provides power control and CPCH control commands (e.g., emergency stop) for the uplink CPCH. The CPCH is also characterized by initial collision and contention. It supports inner-loop power control (fast power control) as opposed to the RACH that relies only on the open-loop power control to transmit the random access request.

3.4.3 Physical Channels

In order to carry the information contained in the transport channels, the physical channels adapt these transport channels (in terms of coding and flows) to the physical layer procedures.

The physical channels consist of frames and slots with a basic radio frame period of 10 ms consisting of 15 slots. The number of bits per slot depends on the physical channel used (spreading factor, modulation, coding rate, etc.). Each physical channel has a specific structure and specific spreading factor (SF) according to the number of bits transmitted on the channel and importance of the information (degree of protection, coding, spreading, etc.).

The main physical channels, specified in Release 99 [18], are: the synchronization channel (SCH), the common pilot channel (CPICH), the acquisition indication channel (AICH), the CPCH status indication channel (CSICH), the paging indicator channel (PICH), the collision detection/Channel assignment indication channel (CD/CA-ICH), the dedicated physical data channel (DPDCH), the dedicated physical control channel (DPCCH), the physical random access channel (PRACH), the primary common control physical channel (PCCPCH), the secondary common control physical channel (SCCPCH), the physical downlink shared channel (PDSCH), and the physical common packet channel (PCPCH).

In Release 5, three physical channels have been added to the 3GPP specifications: the high-speed physical downlink shared channel (HS-PDSCH) intended to carry the HS-DSCH transport channel and two other associated channels intended to carry the related physical control information, the high-speed shared control channel (HS-SCCH) and the high-speed dedicated physical control channel (HS-DPCCH).

In Release 6, the E-DCH dedicated physical data channel (E-DPDCH) has been introduced to carry the E-DCH transport channel. The physical related control information of this channel is carried on the E-DCH dedicated physical control channel (E-DPCCH), the E-DCH absolute grant channel (E-AGCH), the E-DCH relative grant channel (E-RGCH), and the E-DCH hybrid ARQ indicator channel (E-HICH). In addition, a new physical channel, called MBMS notification indicator channel (MICH), has been specified to support the introduction of new high-rate MBMS. The MICH is always associated with the S-CCPCH to which the FACH transport channel is mapped.

Figure 3.4 illustrates the mapping of the transport channels onto the physical channels [18]. This figure shows that only P-CCPCH, S-CCPCH, DPDCH, PDSCH, HS-PDSCH, E-DPDCH, PCPCH, and PRACH can carry higher-layer information (data or signaling) since they correspond to transport channels. The other physical channels do not correspond to any transport channel. These channels are intended to carry control information related to the physical layer only. They are unknown and completely transparent to higher layers. They are exclusively used by the physical layer to harness and control the radio link.

The channels used usually for data transmission are the DPDCH, the HS-PDSCH, and the E-DPDCH. The PCPCH and the S-CCPCH can transmit data in certain cases (e.g., broadcast/multicast services on the S-CCPCH). This book focuses on the interaction between TCP layer and 3G UMTS wireless system. The study of this interaction is more significant in the downlink than in the uplink for the following reasons:

■ The majority of applications carried over TCP is of type context-to-person where the transmitter is a remote host.
■ For uplink services, reduction of the interaction of TCP with the wireless link can be achieved by implementing changes of the TCP protocol at the transmitter (i.e., the mobile station).

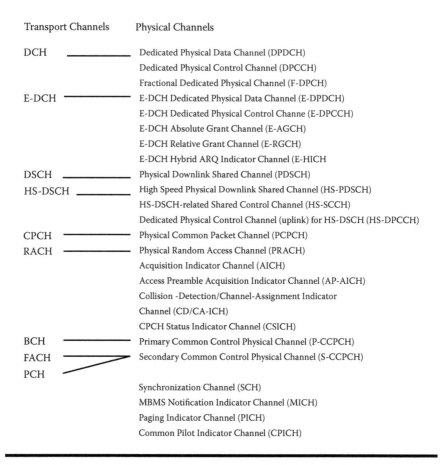

Transport Channels   Physical Channels

DCH — Dedicated Physical Data Channel (DPDCH)
Dedicated Physical Control Channel (DPCCH)
Fractional Dedicated Physical Channel (F-DPCH)
E-DCH — E-DCH Dedicated Physical Data Channel (E-DPDCH)
E-DCH Dedicated Physical Control Channe (E-DPCCH)
E-DCH Absolute Grant Channel (E-AGCH)
E-DCH Relative Grant Channel (E-RGCH)
E-DCH Hybrid ARQ Indicator Channel (E-HICH
DSCH — Physical Downlink Shared Channel (PDSCH)
HS-DSCH — High Speed Physical Downlink Shared Channel (HS-PDSCH)
HS-DSCH-related Shared Control Channel (HS-SCCH)
Dedicated Physical Control Channel (uplink) for HS-DSCH (HS-DPCCH)
CPCH — Physical Common Packet Channel (PCPCH)
RACH — Physical Random Access Channel (PRACH)
Acquisition Indicator Channel (AICH)
Access Preamble Acquisition Indicator Channel (AP-AICH)
Collision -Detection/Channel-Assignment Indicator
Channel (CD/CA-ICH)
CPCH Status Indicator Channel (CSICH)
BCH — Primary Common Control Physical Channel (P-CCPCH)
FACH — Secondary Common Control Physical Channel (S-CCPCH)
PCH
Synchronization Channel (SCH)
MBMS Notification Indicator Channel (MICH)
Paging Indicator Channel (PICH)
Common Pilot Indicator Channel (CPICH)

Figure 3.4   Mapping of transport channels onto physical channels.

Note that in certain cases (e.g., limited bit rate on the uplink), the up-link channel transmitting the TCP acknowledgments can affect the TCP performance on the downlink. This problem can be encountered by implementing some changes (e.g., delaying or transmitting more quickly the TCP acknowledgments) of the TCP receiver entity (i.e., mobile station).

Consequently, in this book, particular attention is given to the DPDCH and HS-PDSCH downlink data channels. The DPCH channel is described in the next section whereas the HS-PDSCH channel is described in the next chapter. Note that detailed structures of all UMTS channels are presented in [18].

3.4.3.1   Dedicated Physical Channel

The DPCH consists of two channels: the DPDCH and the DPCCH. The DPDCH is used to carry the dedicated channel transport channel.

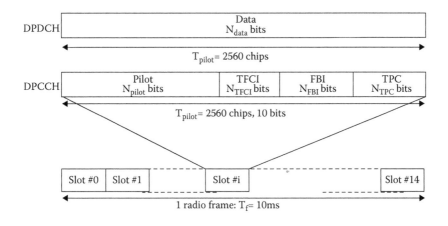

Figure 3.5 Frame structure of the uplink DPCH channel.

The DPCCH is used to carry control information generated at layer 1. The layer 1 control information consists of known pilot bits to support channel estimation for coherent detection, transmit power-control (TPC) commands, feedback information (FBI), and an optional transport-format combination indicator (TFCI). The transport-format combination indicator informs the receiver about the instantaneous transport format combination of the transport channels mapped to the simultaneously transmitted DPDCH radio frame.

In the uplink, the DPCCH and the DPDCH are transmitted in parallel using one spreading code for each one. However, in the downlink these two channels are multiplexed in time and are transmitted using the same spreading code.

Figures 3.5 and 3.6 show the frame structure of the DPDCH and the DPCCH, respectively, in the uplink and the downlink [18]. Each radio

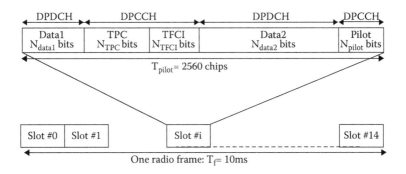

Figure 3.6 Frame structure of the downlink DPCH channel.

frame of 10 ms is split into 15 slots, each of length Tslot = 2560 chips (since the spreading chip rate is 3.84 Mcps). This physical channel supports inner-loop power control at frequency of 1.5 KHz that corresponds to a fast power control period of one slot. The DPDCH spreading factor varies from 256 down to 4 according to the application data rate. The spreading factor of the DPCCH (control) is always equal to 256 resulting in 2560 chips per DPCCH slot (i.e., there are 10 bits per DPCCH slot). More details on these two channels and other physical channels can be found in [18].

3.5 Physical Layer

The physical layer of the UTRAN, in particular the Uu interface, relies on wideband CDMA (WCDMA) described in Chapter 2. The physical layer offers services to the upper layer (MAC layer) via the transport channels. The services offered by the physical layer includes multiplexing/demultiplexing of transport channels and mapping of coded composite transport channels onto physical channels [19,20]. The physical layer handles in addition encoding/decoding of transport channels using a forward error correcting (FEC) code in order to protect the information transmitted over the radio interface against errors. Even though FEC is used, the received signals still contain errors due to severe radio conditions in general. The erroneous packets are retransmitted at the RLC layer using ARQ protocol described in Section 3.11.

In order to achieve a target QoS, i.e., target BER or target signal-to-noise ratio (SNR), the physical layer supports measurements of received signal quality via a pure physical channel called common pilot indicator channel (CPICH). The channel quality is measured in terms of frame error rate (FER), signal-to-interference ratio (SIR), interference power, transmission power, etc. Based on these measurements, the physical layer supports closed-loop power control that adapts the transmission power of physical channels to the short-term radio channel variations in order to achieve a given target QoS for the transmitted information over the radio interface (more details in Section 3.12).

The physical layer supports also soft handover and macrodiversity combining over dedicated physical channels. This is not applied on shared channels. By combining two or more signals, containing the same information and transmitted via two or more node Bs, channel diversity is achieved and signal quality improved at the receiver. The soft handover allows uninterrupted transmission of mobile calls and data sessions, which in turn improves the QoS during handover (see Section 3.13).

In addition to services described above, the physical layer performs other basic operations including [19,20]:

- Radio frequency (RF) processing
- Rate matching
- Modulation and spreading of physical channels
- Demodulation and despreading of physical channels
- Frequency and time synchronization (chip, bit, slot, frame)
- Power weighting and combining of physical channels
- Synchronization and timing advance on uplink channels (TDD only)

3.6 Medium Access Control

The MAC sublayer provides services to the RLC sublayer via logical channels (control and traffic) and coordinates access to the physical layer by mapping these logical channels onto the transport channels. Information at the RLC sublayer, bundled into packet data units (PDUs), is multiplexed by the MAC sublayer into transport blocks and delivered to the physical layer. This multiplexing function allows the mapping of several RLC instances onto the same transport channel, in other words it supports the multiplexing of several logical channels into the same transport channel as we have shown in Section 3.4. Note that on the receiver side, the MAC transport blocks delivered from the physical layer are demultiplexed into RLC PDUs, using the multiplexing identification contained in the MAC protocol control information.

During this mapping, the MAC sublayer performs control of transport formats by assigning the appropriate format for each transport channel depending on the instantaneous source rate in order to achieve efficient use of transport resources. When several logical channels belonging to different users are transported by the same common channel (e.g., FACH, RACH), the user equipment identification (i.e., cell radio network temporary identity [C-RNTI] or UTRAN radio network temporary identity [U-RNTI]) present in the MAC header is used to identify the logical channels of each user on the receiver side.

Priorities between different data flows of one user or between different users sent over common, shared, and E-DCH transport channels can be handled also by the MAC sublayer. Priority between data flows of the same user can be performed by assigning adequate transport formats to each flow so that high-priority flows can be transferred over layer 1 with high bit rate and low priority flows with low bit rate. On the uplink, RACH and CPCH resources (i.e., access slots and preamble signatures for UMTS FDD) are organized in different access service classes (ASCs) (up to 8 ASCs are specified in [16,17]). The MAC sublayer is responsible for applying and indicating to the physical layer the ASC partition associated to a given MAC protocol data unit (PDU). This function provides different priorities on RACH and CPCH [16].

The MAC sublayer is also involved in traffic measurement and monitoring. The amount of RLC PDUs corresponding to a given transport channel is compared to a threshold specified by the RRC layer. The traffic volume measurement is reported to the RRC layer in order to handle reconfiguration or transport channel switching decisions. Ciphering for transparent RLC mode (see next section for transparent mode definition) and execution of the switching between common and dedicated transport channels (decided by RRC) are also performed by the MAC sublayer.

In Release 5, a new MAC entity, called the MAC-high speed (MAC-hs) and located in the node B, has been introduced in the 3GPP specifications. This entity is responsible for handling and managing the HARQ mechanism introduced in HSDPA. The MAC-hs entity as well as the HARQ technique are described in details in the next chapter. The MAC-hs entity is responsible for assembling, disassembling, and reordering higher layer PDUs. The PDUs are delivered in sequence to higher layers. Proper delivery of packets to TCP, the number of uncorrected errors and the experienced delays determine the overall system performance. Chapters 4–7 address the interactions of the MAC and the RLC with the TCP.

3.6.1 MAC Architecture

In order to handle the functions described previously, the MAC layer is divided into the following domains or entities [17]:

- MAC-b is the entity that handles the BCH channel. There is only one MAC-b in each user equipment and one in the UTRAN (node B) as specified by the 3GPP.
- MAC-d is the entity that handles the DCH channel. This entity is specific to each user. In the UTRAN, this entity is located in the SRNC. Note that ciphering is performed by this entity.
- MAC-c/sh/m is the entity that handles the FACH, PCH, RACH, CPCH, DSCH (TDD only) and USCH (TDD only) including ASC selection, transport formats selection, scheduling/priority handling, etc. There is one MAC-c/sh/m located in the user equipment and one in the UTRAN (located in the CRNC). Note that in Release 99, this entity was named MAC-c/sh. In Release 6, it is named MAC-c/sh/m since it is involved in the MBMS services (see Section 3.9) by multiplexing and reading the MBMS Id (which is used to distinguish between MBMS services).

- MAC-hs is the entity that handles the HS-DSCH channel specified in Release 5. This entity, located in the node B and in the user equipment, is responsible for HARQ functionality, transport format selection and scheduling.
- MAC-m is the entity that controls access to the FACH channel when it is used to carry MTCH and MSCH logical channels. This entity is added to the 3GPP specifications in Release 6. It exists only in the user equipment side of the MAC architecture in the case of selective combining of MTCH channels from multiple cells.
- MAC-e/es are the entities that handle the E-DCH channel. These entities are introduced by the 3GPP in release 6.

The general MAC architectures on the user equipment and the UTRAN sides are, respectively, depicted in Figures 3.7 and 3.8 [17].

3.6.2 Protocol Data Unit

Peer-to-peer communication is achieved by the exchange of PDUs. The MAC PDU presented in Figure 3.9 [17], consists of a MAC header and a MAC service data unit (SDU) both of variable size. The content and the size of the MAC header depend on the type of the logical channel, and in some cases none of the parameters in the MAC header are needed. The size of the MAC-SDU depends on the size of the RLC PDU, which is defined during the set-up procedure.

In the 3GPP specifications [17], the MAC PDU structure is described in details for all transport channels. Since this chapter focuses on UMTS Release 99, only the MAC PDUs of common, shared, and dedicated channels except HS-DSCH and E-DCH are considered. Note that the MAC-d PDUs of HS-DSCH and E-DCH are similar to the MAC-PDU described in this section. Only, the MAC-hs and MAC-e/es PDUs are distinct when HS-DSCH and E-DCH are used. The MAC PDU header contains optionally the following fields [17]:

- The target channel type field (TCTF) is a flag that identifies logical channels such as BCCH, CCCH, CTCH, SHCCH, MCCH, MTCH, MSCH carried on FACH, USCH (TDD only), DSCH (TDD only) and RACH transport channels. Note that the size of the TCTF field of FACH for FDD is two, four, or eight bits.
- The C/T field is used to identify the logical channel instance and type carried on dedicated transport channels and potentially on the FACH and RACH channels (only when they are used for user data transmission). The identification of the logical channel instances is

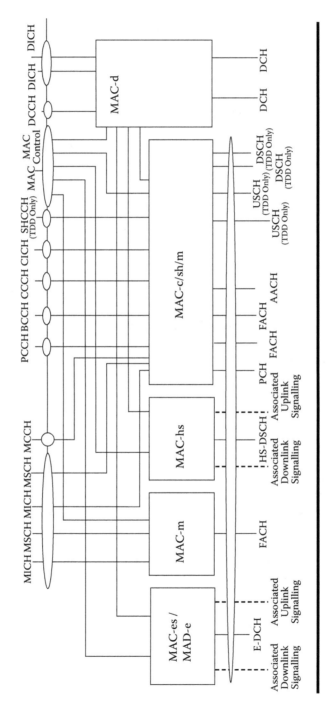

Figure 3.7   General MAC architecture of the UMTS Release 6 on the UE side.

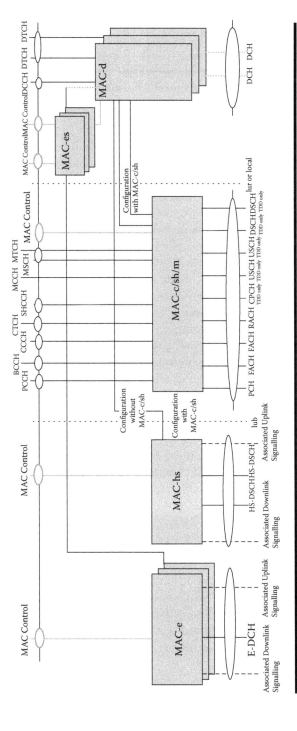

Figure 3.8 General MAC architecture of the UMTS Release 6 on the UTRAN side.

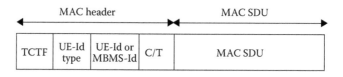

Figure 3.9 Structure of a MAC protocol data unit (PDU).

mandatory when multiple logical channels are carried on the same transport channel. The size of the C/T field is fixed to four bits for both common transport channels and dedicated transport channels.

- The user equipment identity type field provides identification of the user equipment on common transport channels. Two types of user equipment identities are defined by the 3GPP: (1) UTRAN radio network temporary identity (U-RNTI) is used only in the downlink direction (never in the uplink) when DCCH logical channel of RLC unacknowledged mode (UM) is mapped onto common transport channels. The RLC UM mode is described in Section 3.7 of this chapter; (2) cell radio network temporary identity (C-RNTI) is used for DTCH and DCCH in uplink, DTCH (and maybe DCCH) in downlink, when mapped onto common transport channels.
- The user equipment identity type field helps the receiver to correctly decode the user equipment identity in MAC headers.
- The MBMS-identity, being added to the specifications in Release 6, is used only in the downlink to provide identification of MTCH for an MBMS service carried on FACH transport channel.

As we have indicated above, these MAC header fields are used optionally (i.e., depending on the logical channel and the transport channel on which this logical channel is mapped). To explain this further, Figure 3.10 provides an example of MAC PDU header when DTCH or DCCH are mapped onto dedicated, common or shared transport channels (except E-DCH and HS-DSCH) [17]. In this figure, five cases are considered:

(1) DTCH or DCCH is mapped onto DCH. In this case, none of the MAC header field is used since no multiplexing of dedicated channels at the MAC sublayer is considered.

(2) DTCH or DCCH is mapped onto DCH. In this case, the C/T field is included in the MAC header since multiplexing of dedicated channels on MAC is considered.

(3) DTCH or DCCH is mapped onto RACH/FACH. In this case, the fields TCTF, C/T, user equipment identity type and user equipment identity are included in the MAC header. For FACH, the user equipment

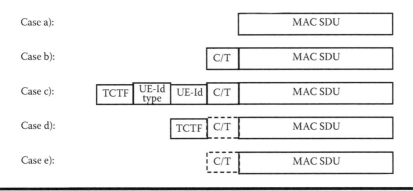

Figure 3.10 Example of a MAC PDU header structure.

identity field used can be either the C-RNTI or U-RNTI whereas for RACH only the C-RNTI is used.

(4) DTCH or DCCH is mapped onto DSCH or USCH for UMTS TDD only. In this case, the TCTF field is included in the MAC header. In addition, the C/T field may be used if multiplexing on MAC is applied. Note that this case exists in TDD mode only (since DSCH and USCH are used for TDD mode).

(5) DTCH or DCCH is mapped onto DSCH or USCH where DTCH or DCCH are the only logical channels. Only the C/T field may be included in the MAC header if multiplexing on MAC is applied. Note that this case exists in TDD mode only (since DSCH and USCH are used for TDD mode).

3.7 Radio Link Control

The RLC [21] sublayer provides radio link services for use between the user equipment and the network. The RLC sublayer contains discrete RLC entities required to create radio bearers. For each radio bearer, the RLC instance is configured by RRC [22] to operate in one of three modes: transparent mode (TM), unacknowledged mode (UM), or acknowledged mode (AM). The modes are used according to the application characteristics.

The upper layer PDUs are delivered to the RLC sublayer and fit into RLC SDUs. Each RLC SDUs is then segmented in one or more RLC PDUs, which are mapped onto MAC PDUs. Examples of data flow mapping between RLC and MAC layers for transparent and nontransparent (UM or AM) modes are depicted, respectively, in Figures 3.11 and 3.12. The two types of RLC PDUs are defined in the 3GPP specifications [21]. The first corresponds to PDU or data-PDU used to carry the upper layers information (data

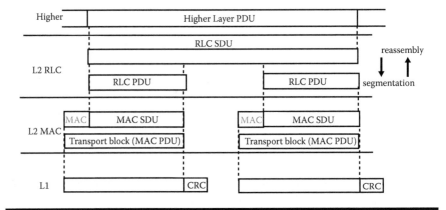

Figure 3.11 **Example of data flow mapping between RLC transparent mode and MAC.**

or signaling). The second is the status-PDU used to carry the related RLC control information. The detailed structures of these PDUs are given in [21].

3.7.1 Transparent Mode (TM)

TM is used in general for time-critical data such as speech services. Since no signaling between RLC entities and RLC PDUs is specified by the 3GPP for this mode, TM is not capable of detecting PDUs lost in transmission from the peer RLC entity. In this mode, the RLC layer performs segmentation of upper layer PDUs into RLC PDUs and vice-versa (i.e., assembly of RLC PDUs

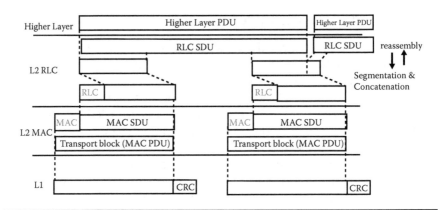

Figure 3.12 **Example of data flow mapping between RLC nontransparent mode (UM or AM) and MAC.**

into variable length higher layer PDUs). In addition, this layer contains a discard timer allowing the RLC transmitter to discharge a given RLC SDU from the buffer if the timer has expired. The SDU discard function will be described in more detail later in this section. To summarize, the RLC layer performs the following functions [21]:

- Segmentation/reassembly of the upper layer PDUs
- Transfer of user data
- SDU discard

3.7.2 Unacknowledged Mode (UM)

In unacknowledged mode (UM), RLC signaling is added to the data. A sequence number is assigned to each PDU so that the UM RLC can indicate the PDUs lost in transmission from the peer UM entity. A lost PDU is detected by receiving a PDU with an out of order sequence number. This error detection function is called sequence number check. At the receiver side, the RLC entity can deliver the recovered RLC SDUs to the higher layer using two strategies:

1. Immediate delivery of the recovered SDUs even if the previous SDUs are not recovered correctly, which generates out-of-sequence data at the higher layer. This strategy is suitable for time critical data services.
2. In-sequence delivery of the SDUs to the higher layer by reordering the received SDUs at the receiving entity of the RLC sublayer. This strategy results in additional delays and is more suitable for nontime critical services.

Note that in both strategies, erroneous SDUs are discarded and this results in missed SDUs at the higher layer. This mode is only suitable for services with low residual SDU error ratio requirement. Note that missed SDUs can be generated by discharging SDUs at the RLC transmitter buffer due to timer expiration. The unacknowledged mode is essentially used for services carried over the UDP protocol and not TCP. Missing data at the transport layer (due to missed erroneous SDUs) may be misinterpreted by TCP as congestion. TCP would trigger unwanted retransmissions and congestion window control. This undesired interaction of TCP with the RLC layer in wireless systems will be explained in more details in Chapters 5, 6, and 7.

In order to take maximum advantage of the variable radio channel capacity (which requires variable MAC PDU sizes), UM provides the possibility to segment or concatenate data from SDUs to fit various sized MAC PDUs. Padding bits can be used to fill a RLC PDU when concatenation is not applicable and the remaining data to be transmitted does not fill an entire RLC PDU. The RLC UM mode is characterized by the following RLC functions [21]:

- Segmentation and reassembly
- Concatenation
- Padding
- Transfer of user data
- Ciphering to avoid unauthorized data acquisition
- Sequence number check
- SDU discard
- Out-of-sequence SDU delivery
- Duplicate avoidance and reordering

3.7.3 Acknowledged Mode (AM)

In acknowledged mode (AM), additional PDUs are defined to allow bidirectional signaling between peer RLC entities. These additional PDUs are used to request the retransmission of missing or erroneous data at the receiver. In TM and UM, the erroneous PDUs are discarded. In the case of AM, a STATUS–REPORT is sent to the peer RLC entity asking it to retransmit the erroneous PDUs using the ARQ protocol (see Section 3.10 for more details on ARQ). Consequently, this mode is suitable for nontime critical services with high quality/integrity requirements carried over TCP or UDP transport protocols (but essentially TCP). The recovered SDUs are delivered to the upper layer in sequence or out of order. In Chapters 5–7, we will see that in sequence delivery of SDUs results in better performance at the TCP layer since out-of-sequence reception of packets may cause a triple duplicate phenomenon (misinterpreted as congestion by TCP), which results in packet retransmissions and congestion window downsizing at the TCP layer. The functions handled by the RLC sublayer in the AM mode can be summarized as follows [21]:

- Segmentation and reassembly.
- Concatenation.
- Padding.
- Transfer of user data.
- Ciphering to avoid unauthorized data acquisition.
- Error correction (using ARQ).
- SDU discard.
- In-sequence SDU delivery.
- Duplicate detection used when the RLC receiver detects the reception of the same RLC PDU more than one time. The RLC entity delivers the resultant upper layer PDU only once to the upper layer.

- Flow control to control the transmission rate between the peer RLC entities.
- Protocol error detection and recovery.

3.7.4 SDU Discard at the RLC Sender

In the 3GPP specifications [21], the RLC sender entity can be configured by upper layers (in particular RRC layer) to discard SDUs from the sender buffer in certain cases. The objective of this function is to manage the QoS requirements, in terms of delays and error ratio, of the service application carried over the UMTS system.

The reasons to discard SDUs at the RLC sender change from an RLC mode to another. In all RLC modes (TM, UM, and AM), once the RLC layer receives a SDU from the upper layer (PDCP sublayer), a timer specific to this SDU is started. If the timer expires before delivering the corresponding SDU to the MAC layer (i.e., before transmitting the SDU from the sender), the corresponding SDU is discarded. The presence of the timer in the RLC sender entity delimits the transfer delay of the SDUs to the receiver at the expense of increased SDU loss ratio. The RRC layer configures the RLC timer, as well as the radio bearer between the UE and UTRAN to meet the QoS requirements (i.e., SDU loss ratio and SDU transfer delay). Note that depending on the RLC modes used, two timer discard modes can be distinguished: (1) timer-based discard without explicit signaling used in UM and (2) TM modes and timer-based discard with explicit signaling used in the AM mode.

In the RLC AM mode, in addition to timer-based discard, SDUs discard may be generated when the maximum number of a given PDU retransmission is reached. In certain cases, severe wireless channel conditions may result in successive retransmission failure (i.e., erroneous) of the same PDU. The RLC discards the corresponding SDU when the number of retransmission reaches a maximum, noted MaxDAT, to limit the SDU transfer delay and meet QoS requirements. For services with no delay restrictions (e.g., background applications class), the SDU is not discarded even if the maximum number of the corresponding PDU retransmissions (MaxDAT) is reached. Four SDU discard modes have been specified by the 3GPP in [21]:

- Timer-based discard with explicit signaling applicable to RLC AM mode
- Timer-based discard without explicit signaling applicable to RLC TM and UM modes
- SDU discard after MaxDAT transmissions applicable to RLC AM mode
- No discard after MaxDAT transmissions applicable to RLC AM mode

3.7.4.1 Timer-Based Discard with Explicit Signaling

This discard mode is only applicable for RLC AM mode. For every delivered SDU from the upper layer (PDCP sublayer) a timer is started to monitor the lifetime of the SDU in the RLC sender buffer. The values of the timer can range from 100 ms up to 7.5 sec. If the timer of a given SDU expires, the corresponding SDU is then discarded from the buffer. The RLC sender notifies explicitly the receiver about this SDU discard event using the field, Super Field More Receiving Window (SuFi MRW), contained in the so-called status-PDU. The receiver acknowledges the receipt of this super field via the SuFi MRW-ACK sent to the sender and advances its receiving window to skip the missing PDU.

When the SuFi MRW field is transmitted, a timer is started in the RLC sender entity. If the timer expires before receiving the SuFi MRW-ACK from the receiver (e.g., the SuFi MRW gets lost), the SuFi MRW is retransmitted and the timer is reinitialized. Note that the timer expires after a time ranging between 50 ms and 900 ms according to the configuration signaled by the RRC layer. If the number of transmissions of SuFi MRW reaches the maximum number MaxMRW (configured by RRC in a range from 1–32), the explicit SDU discard procedure is stopped and a reset procedure is initiated by the RLC sender.

The reset procedure consists of transmitting a reset PDU, initializing a specific timer and expecting reception of a reset PDU-ACK from the receiver before the timer expires. In case of timer timeout, the reset PDU is retransmitted and the timer is reinitialized. The timer timeout can range from 50 ms up to 1 sec and the maximum number of reset PDU transmissions lies in the range 1–32. If the maximum number of reset PDU transmissions is reached, the RLC signals to the RRC an unrecoverable error and a failure of Reset procedure. The RRC layer can decide then to reconfigure or to release the corresponding radio bearer.

3.7.4.2 Timer-Based Discard without Explicit Signaling

This timer-based discard of SDUs is applicable for RLC TM and UM modes. In this mode, timer expiration results in discarding the corresponding SDU like the previous discard mode. However, the sender does not inform explicitly the receiver about this SDU discard event. The sender continues the transmission of the subsequent SDUs. In the UM mode, the receiver deduces implicitly the discard event by receiving out of sequence SDUs (since the sequence number check function is used in this mode). In the TM mode, no error detection in the received PDUs is used. The RLC receiver entity delivers merely the received data to the upper layer (transfer of user data function cited above).

3.7.4.3 SDU Discard after MaxDAT Transmissions

In the RLC AM mode, the use of ARQ results in retransmitting the erroneous PDUs to deliver error free data to the upper layer. If the number of transmissions of a given PDU reaches the maximum MaxDAT, configured by the RRC layer, this PDU and the other PDUs of the same SDU (since a SDU may contain more than one PDU) are discarded. As in the timer-based discard with explicit signaling, the RLC sender notifies explicitly the receiver about this discard event using the field SuFi MRW contained in the corresponding status-PDU. The same explicit signaling and subsequent reset procedures, described above in the timer-based discard, are used in this SDU discard mode. Note that the parameter MaxDAT, configured by RRC, can range from 1–40.

3.7.4.4 No Discard after MaxDAT Transmissions

This mode is similar to the "SDU discard after MaxDAT transmissions mode," described above, except for when the number of transmissions MaxDAT is reached. The explicit signaling procedure is skipped and the reset procedure is initiated immediately.

3.8 Packet Data Convergence Protocol (PDCP)

The PDCP sublayer [23] belongs to the data link layer in the user plane and for the switched domain only. The PDCP allows the use of network, transport, and upper layer protocols like TCP/IP or UDP/IP over the UTRAN by converting the network packets into RLC Service Data Units (SDUs) and vice versa.

In addition, this sublayer provides header compression and decompression of the IP packets conveyed in the UTRAN. Since only few TCP/IP header fields change from IP packet to another and the majority of header fields remain intact, the PDCP can support the compression of subsequent IP packets, which results in more efficient use of the radio resources. In the 3GPP specifications [23], several header compression algorithms are presented (in particular those proposed in [1,24].

The PDCP supports also reliability of data transfer in the UTRAN. In certain cases (e.g., SRNS relocation), the RLC does not provide reliability for data transmission. In addition, when upper layers (e.g., transport layer) are not reliable (e.g., UDP) and that reliability is achieved at the application layer (e.g., streaming), errors caused by SRNS relocation cannot be recovered by retransmission of erroneous packets at the transport layer. In this case, by maintaining the same PDCP sequence number, the PDCP sublayer avoids losses and provides reliability for these data services.

3.9 Broadcast/Multicast Control (BMC) and Multimedia Broadcast/Multicast Service (MBMS)

The Broadcast/Multicast Control (BMC), described in [25] and active only in the user plane, supports and control cell broadcast service (CBS) over the UMTS radio interface. The SMS cell broadcast service is the only service, used in Release 99, that utilizes this protocol sublayer. The SMS CBS service relies on the unacknowledged mode (UM) of the RLC and uses the logical common traffic channel (CTCH), which is mapped into the transport forward access channel (FACH).

The BMC is the entity that stores the cell broadcast messages (associated with scheduling information) received over the interface between the cell broadcast center (CBC) and the RNC. On the UTRAN side, the BMC estimates the appropriate transmission rate of the broadcast service over the FACH channel and requests the required resources from the RRC. In addition, the BMC generates and transmits, to the cell users, schedule messages that indicate the radio frame that should contain the transmitted CBS messages. In the user equipment, the BMC sublayer receives the schedule messages and determines the appropriate radio frame containing the CBS messages. This information is then transferred to the RRC responsible for management and configuration of the physical layer for discontinuous reception. Consequently, only the radio frame containing the CBS messages is received by the cell users. In the UE, the BMC delivers the error-free CBS messages to the upper layers.

Basically, the CBS is used for low data rate services (SMS cell broadcast). In Release 6 [26], the MBMS is introduced to convey higher rate broadcast/multicast information over the radio interface (e.g., 64 kbps). According to the number of cell users, the system could decide to transmit the messages via MBMS using point-to-point or point-to-multipoint transmission. When point-to-point transmission is used, the MBMS content is transmitted over the dedicated transport channel (DCH) of each user (that should receive this service). In this case, the logical channel can be the DCCH (for related control information) or the DTCH (for user data) and the DCH transport channel is mapped into the DPDCH physical channel. When the MBMS content is transmitted using point-to-multipoint transmission, two new logical channels are being specified in Release 6 to carry the MBMS information: MBMS point-to-multipoint control channel (MCCH) to carry control information and MBMS point-to-multipoint traffic channel (MTCH) to transport user data. These two logical channels are mapped onto the FACH transport channel, which in turn mapped onto the secondary common control physical channel (S-CCPCH).

3.10   Radio Resource Control (RRC)

The RRC [22] layer is the most complex layer in the UTRAN. It is involved in the management of connection between the user equipments and RNC by handling the control plane signaling of layer 3 between the user equipments and UTRAN.

The principal function of RRC is the establishment, reconfiguration and release of radio bearers, transport channels and physical channels, and this on request from higher layers. Establishment and reconfiguration consists of performing admission control and selection of parameters allowing to describe the processing of radio bearers in layers 1 and 2. Note that a radio bearer is an association of functions at various levels to handle information transmission (signaling and nonsignaling) between the user equipment and UTRAN. On request from higher layers on the user equipment side to establish or release user equipment signaling connection, the RRC performs establishment, management, and release of RRC connections. Note that RRC connection release can be generated by a request from higher layer or by the RRC layer itself in case of RRC connection failure. In addition, the RRC layer is involved in the user equipment measurement reporting including control of the parameters to be measured (e.g., air interface quality, traffic) and the period of measurements as well as the report format. Based on these measurements, the RRC performs control of radio resources in both uplink and downlink including coordination of radio resources allocation between the different radio bearers of the same RRC connection, open-loop power control to set the target of the closed-loop power control, allocation of sufficient radio resources to achieve the radio bearers QoS, etc. The RRC keeps track of the user equipment location and handles mobility functions for the RRC connection allowing handover mechanism including inter- and intrafrequency hard handover, intersystem handover and intersystem cell reselection (intersystem means UTRAN and another system, e.g., GPRS).

Besides, the RRC layer handles system information broadcasting from the network to all users or to a specific group of users. It is involved in the broadcasting of paging information initiated by higher layers or the RRC layer itself during the establishment of the RRC connection. This layer controls the BMC sublayer and MBMS services (see previous section). Depending on the traffic requirements of the BMC sublayer, the RRC performs initial configuration and radio resource allocation of the CBS (e.g., mapping schedule of CTCH onto FACH) and configure the physical layer at the user equipment for discontinuous reception. Consequently, only the radio frame containing the CBS messages is received by the cell users.

In addition to the functions described above, the RRC performs other procedures such as ciphering control (i.e., on/off) between the user equipment and UTRAN, routing of higher layers PDUs to the appropriate higher layer entity and initial cell selection based on measurements in idle mode and selection procedures.

3.11 Automatic Repeat Request Protocol

The information transmitted over the air interface is protected against errors by the use of the FEC code called the channel code. In addition, to counteract the fast fading effect, some averaging techniques such as long interleaving, wideband spread spectrum, and frequency hopping are used to average the effect of the fading over all the transmission time so that temporary bad channel conditions are compensated by short-term good channel conditions. These techniques use fixed parameter values to deal with the worst channel conditions. In other words, channel coding rate, interleaving, and spreading bandwidth, are all fixed during the system conception phase. In UMTS, as in the majority of wireless systems, the adaptation to the short-term channel variations is performed using fast power control.

When services, either voice or data, are transmitted over wireless systems, required QoS constraints must be satisfied. The service QoS is generally characterized by a delay constraint and a tolerance to errors. For voice service, the BER should not exceed 10^{-3}, but no delay is tolerated. For nonreal-time data services (i.e., interactive and background classes) the BER should be less than 10^{-8}. Some delay in receiving NRT services is usually acceptable. To achieve a target BER of 10^{-8}, two possibilities can be envisaged: increasing the transmitted power at the node B or increasing the redundancy in channel coding. The first solution results in a drastic decrease of cell capacity since more power from the total and limited available node B power of 43 dBm should be allocated to each user. In this case, the interference induced by a user on other users would also increase, thereby decreasing the cell capacity even more. The second solution lowers the achievable user bit rate since more header or redundancy is added to the information sequences. Consequently, both solutions induce a capacity loss. To alleviate this degradation, the ARQ protocol has been proposed and widely used in current wireless systems to achieve error-free data delivery over the air interface for NRT applications.

The ARQ protocol consists of retransmitting erroneously received information until error-free reception of the information packets occurs. Since NRT data tolerates certain delays, the idea operates at much higher BER (e.g., 10^{-3} instead of 10^{-8}) and compensates for the increased error rates by retransmitting erroneously received packets until error-free transmission is achieved. Solutions are developed in wireless system: The target BER is increased to a higher value (e.g., 10^{-3}), which causes an increase of errors in the received information. This solution results in savings of resources through operation at higher BER at the expense of increased delays in packet reception. A trade-off should be found between capacity improvement and increased delays. The incurred delays over the radio interface can also have a drastic effect on overall end-to-end system performance since the TCP is often used for NRT applications in the fixed networks and the Internet to manage congestion control. The ARQ protocols and TCP can negatively interact and can lead to effective system throughput and capacity degradation. This would defeat the original purpose of improving cell capacity in the radio access networks and the end-to-end applications throughput as well. Such conflicts, where TCP mistakenly takes retransmissions over the air by the ARQ protocol for congestion in the fixed network segments, are addressed in Chapters 6 and 7. Chapter 6 presents solutions for wireless TCP—link-layer solutions and TCP variants—and Chapter 7 specializes the study to the hybrid ARQ protocol used in the UMTS HSDPA and shows how conflicts with TCP can be avoided using intelligent scheduling.

Now a description of some well-known ARQ protocols is provided. The use of ARQ does not eliminate all errors since the typically used CRC decoders do not detect all errors. The probability of errors at the output of the ARQ can be evaluated using P_{ud}, the probability of undetected errors by the CRC, and P_d, the probability of detected erroneous packets, which is also the probability of retransmission. It is assumed that the feedback receiver to sender is error free. The packet error rate, P_e, is evaluated by summing the probabilities of events resulting in the acceptance of erroneous packets [27]:

$$P_e = \sum_{j=0}^{\infty} P_d^j P_{ud} = \frac{P_{ud}}{1 - P_d}. \tag{3.1}$$

For most of the data network systems three ARQ protocols can be used: the SW protocol, the go-back-n protocol, and the SR protocol [28].

In the UMTS R99 [21], the SR-ARQ is used at the RLC level. In fact, the cyclic redundancy check (CRC) decoder detects the presence of errors in each RLC PDU at the user equipment. In case the RLC-PDU is in error, the user equipment informs the RNC via the node B by transmitting a nonacknowledgment of the PDU over the uplink DPCCH. The RNC retransmits the erroneous RLC-PDU until it is received without errors.

3.11.1 SW Protocol

This is the simplest ARQ protocol. The sender or the RNC in the case of UMTS R99, classes the packets to transmit in a first input first output (FIFO) buffer, transmits the first packet in the buffer to the receiver, starts a timer, and waits for an acknowledgment from the receiver. A nonacknowledgment or a timer expiration causes a retransmission of the same packet by the sender. Once a positive acknowledgment is received before timer expiration at the sender, the next packet in the buffer is transmitted to the receiver. This strategy causes high delays since the packets in the buffer cannot be transmitted before receiving the acknowledgment of the previous packet. The time of inactivity elapsing between the transmission of a packet and the reception of the acknowledgment makes this protocol inefficient [28]. Note that in UMTS R99, this protocol is not used.

3.11.2 Sliding Window Protocol

To deal with the inefficiency problem of the SW strategy, a sliding window protocol was developed. Instead of transmitting a packet and waiting for the acknowledgment before sending another packet, the sender transmits W packets, where W is a transmission window size, before receiving the acknowledgment of the first one. Once the acknowledgment of the first packet is received, another packet is transmitted, so that the total number of the transmitted packets that await an acknowledgment is maintained equal to the window size W. This strategy increases the efficiency of the system but requires larger headers; for example, a sequence number should be attributed to each packet.

If the received packet contains errors, a negative acknowledgment is sent, for example, on the uplink DPCCH to the sender, for example the RNC. In this case, two control strategies can be applied: go-back-n and SR.

The go-back-n control protocol manages several blocks at a time. When a received packet is erroneous, a negative acknowledgment is transmitted to the sender. All transmitted packets starting from the erroneous one have to be retransmitted. At the receiver, all packets received past the erroneous one are discarded. This protocol is used in the TCP (see Chapter 5).

In wireless networks, this protocol is suited only for system carrying high bit rates over the air interface (gigabits/s) since bursty errors generated by multipath channels affect several successive packets in this case. In wireless systems with limited bit rates such as UMTS, this strategy generates additional delays and limits the system efficiency. Selective retransmission of erroneous packets is more appropriate for low data-rate systems.

The inefficiency of the go-back-n protocol in the case of high errors rate can be solved by using the SR strategy. If a received packet is erroneous, only this packet is retransmitted by the sender, which results in decreasing the number of packet retransmissions. If P_d represents the detected packet error rate, the average number of packet transmissions can be evaluated using the following equation:

$$N = \sum_{j=1}^{\infty} j P_d^{j-1}(1 - P_d) = \frac{1}{1 - P_d}. \tag{3.2}$$

This protocol requires more header and more complex receivers. An out-of-sequence problem can arise since some packets can be correctly received and decoded before others. This can have a drastic effect on higher-layer protocols such as TCP. Therefore, in-sequencing of these packets should be performed before their transmission to higher layers (see Chapters 6 and 7).

3.12 Power Control

Power control is one of the key techniques used in wireless networks. As explained in Chapter 1, the radio channel changes instantaneously according to such things as mobile position, environment, scatters, and shadowing. To achieve a required QoS per user, the transmission power must be adapted instantaneously to the channel variations to maintain a given signal level at the receiver. This would maintain a given BER at the receiver. As QoS requirements depend on service type, different target BERs are expected. For speech service, the target BER is 10^{-3}. For data the target BER is about 10^{-8} to achieve the high integrity requirements of data. For a given mobile scenario (e.g., environment, channel type, frequency, mobile speed), the target BER corresponds to a specific value of SIR called target SIR. If the scenario changes, the target SIR must be changed to maintain the same BER. Consequently, power control must provide an instantaneous adaptation of the data transmission to the QoS requirements and the mobile scenario by adapting the target SIR and the instantaneous transmitted power to achieve this target SIR.

In UMTS, this adaptation can be conducted continuously by using the radio link quality measurement when connectivity is already established and before the connection to the RRC that regulates access over the air interface. Consequently, two forms of power control exist in UMTS: open-loop and closed-loop [29,30]. The first one is used on the uplink when users attempt access for the first time and the second during active connections or sessions.

3.12.1 Open-Loop Power Control

Open-loop-power control consists of an estimation of the initial uplink and downlink transmission powers. By receiving the CPICH and the control parameters on the BCCH, the user equipment sets the initial output power, for uplink channels (such as PRACH or the first transmission of the DPCCH before the start of the inner loop), to a specific value to achieve a target SIR at the receiver without generating a high interference level on the other users' signals. In the downlink, the node B on user equipment measurements sets the transmission power (e.g., DPCH first transmission before the inner-loop start, S-CCPCH carrying FACH and PCH) at an adequate level to achieve as much as possible an acceptable signal level without generating much interference. Note that the RNC is not involved in the open-loop power control, since it is a one-way power control between the node B and the user equipment only.

3.12.2 Closed-Loop Power Control

Closed-loop power control is performed continuously during active sessions known as RRC connection and involves three entities: the user equipment, the node B, and the RNC. Two mechanisms called inner loop and outer loop are running among these three entities to adjust the instantaneous power and the target SIR.

The inner-loop power control is the first part of the closed-loop power control performed between the user equipment and the node B. The inner-loop power control, based on SIR measurements at receivers, is used to combat fast channel variations and interference fluctuations. The power-control algorithm aims at keeping user SIR at an appropriate level by adjusting the transmission powers up or down. This power control operates on each slot—that is, at a frequency of 1500 Hz. In case of uplink, each base station in the active set evaluates the received SIR by estimating the received DPCH power after Rake processing of the concerned mobile signal and the total uplink received interference. In case of downlink, the mobile station computes the received SIR by estimating the DPCH power after Rake combining of the received signal and the total downlink interference power received at that mobile station. In both uplink and downlink, TPC commands are generated and should be transmited over each slot according to the following rule: If $SIR estimated > SIR target$ then the TPC command to transmit is 0, whereas if $SIR estimated < SIR target$ then the TPC command to transmit is 1. Note that in the downlink, the TPC command is unique on each slot or can be repeated over 3 slots according to the mode used. In addition, the power step size can equal to 0 in the downlink if a certain "Limited Power Increase Used" parameter is used and the so-called "Power-Raise-Limit" is reached (see [29] for more details). In the uplink, if the TPC command is 1 (or 0) the user equipment might increase (or decrease) the output power of the DPCCH channel by a step size of 1 dB. A step size of 2 dB is also specified to use in certain cases. Since the DPCCH and DPDCH channels are transmitted in parallel using two different codes, they are transmitted at two different power levels depending of the transport format combination (TFC) of these channels. The user equipment estimates a gain factor between the powers of these channels and adjusts the power of the DPDCH according to the DPCCH power. In the downlink, the DPCCH and DPDCH channels are transmitted on the same channel (i.e., same code). In this case, the node B adapts the transmission power by a step size of 0.5, 1, 1.5, or 2 dB according the received TPC command. Note that the support of 1 dB step size

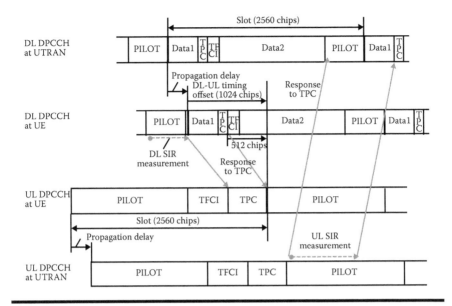

Figure 3.13 Example of the fast power control timing for DPCH channel [29].

by the node B is mandatory while the other step sizes are optional. An example of the fast power-control timing in uplink and downlink is depicted in Figure 3.13. Note that the power control dynamic range is 25 dBm. The maximum allowed output power per channel is 36 dBm, and the maximum base station output power is limited to 43 dBm.

The outer-loop power control maintains the quality of communication for each service by using power as low as possible. The RNC performs outer-loop power control by adjusting the SIR target dynamically to keep the block error rate (BLER) in a given range. The uplink outer loop is located in the RNC. The RNC adjusts the target SIR by receiving the measured SIR from the node B and by measuring the BLER. The downlink outer loop is located in the user equipment. This procedure is slow and can be conducted 10 to 100 times per second (i.e., at a frequency ranging from 10 up to 100 Hz).

3.13 Handover

UMTS system supports the use of hard, soft, and softer handover [30] between cells. The objective of the handover use in cellular system is to achieve an optimum fast closed power control, in other words to allow mobile connection to the strongest cells. The strongest cell is the cell that has the best wireless link conditions to transmit the data to a specific user, i.e., that allocates the lowest power resources to this user though with achieving the required service QoS (target BLER).

- The hard handover is applied when the link quality of a user to an alternative cell becomes better than that to the serving cell. The active connection in this case changes from the serving to the other. This kind of handover is applied in following four cases: handover between GSM and UMTS, intermode handover (FDD/TDD and vice versa), intramode handover across different frequencies (the serving cell and the alternative cell do not have the same frequency), and intrafrequency handover for services using a shared channel.

- In the 3GPP specifications, soft/softer handover can be used for services using a dedicated channel and between cells having the same radio access technology (i.e., CDMA) and the same frequency. Note that softer handover consists of a soft handover between different sectors of the same cell. The basic idea of a soft handover is to allow mobile station connection to more than one node B at the same time. This results in a seamless handover with no disconnections of radio bearers. In addition, the mobile station combines using maximal ratio combining (MRC) the signals received from different node Bs (called macrodiversity). This increases the SIR level and ensures good signal quality received by users at the cell border (where in general insufficient signal level is obtained from a single cell). In the uplink, the macrodiversity can also be applied by combining the signals transmitted by the same mobile and received by different node Bs (different sectors in the case of softer handover). These received signals are combined in the RNC in the case of soft handover or in the base station in the case of softer handover which improves the uplink received SIR and allows a reduction of the mobile transmission power.

A mobile station in CDMA systems is connected to more than one base station (these base stations are called the active set) if the path losses are within a certain handover margin or threshold denoted by h_m or "reporting–range." Handovers due to mobility are performed using a hysteresis mechanism defined by the handover hysteresis h_{hyst} and replace hysteresis (rep_{hyst}) parameters. Figure 3.14 [30] depicts an example of soft handover wherein three cells are considered and ΔT is the time to trigger. A node B is added into the active set if the received pilot signal level (i.e., SIR over the CPICH which reflects the path loss) is greater than the best pilot level (i.e., the best path loss) by more than $h_{hyst} - h_m$ ($pilot_{level} > bestpilot_{level} - h_m + h_{hyst}$). A base station is removed from the active set if $pilot_{level} > bestpilot_{level} - h_m - h_{hyst}$. Similarly, a base station outside the active set can replace an active

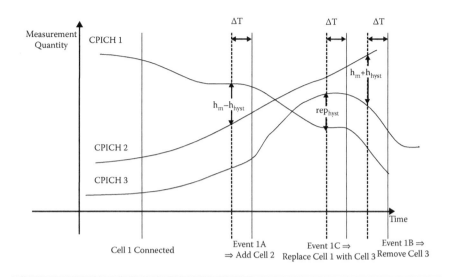

Figure 3.14 Example of soft handover algorithm [30].

base station if its path loss becomes more than rep_{hyst} of that base station. For more details on handover procedures, reader can check reference [30].

3.14 Modeling and Cell Capacity

To predict the efficiency of the UMTS R99 system, an estimation of the cell capacity should be conducted. The cell capacity is the maximum number of users that can be served simultaneously, each having a given bit rate and a given QoS (e.g., given SIR) to satisfy. Since the UMTS, like all CDMA-based systems, offers soft capacity—CDMA systems degrade smoothly as more codes or resources are allocated to users or as more users are accepted in the system—the number of users can exceed the capacity bound at the expense of QoS, in other words receiving the data with a SIR less that the SIR target. To assess this degradation, the notion of *outage probability* was introduced to indicate how much, or how many times, the SIR of the received data does not achieve the target SIR value. The *outageprobability* is used to assess coverage in CDMA systems and to provide the needed flexibility in cell planning that is not possible in TDMA-based systems that exhibit hard capacity. Once all available time slots or channels are allocated in TDMA, the system cannot accept any more users.

In the literature, the cell-planning problem has been studied widely with various analytical and simulation methods [e.g., 31–45]. In this section, a basic model for the cell capacity, developed in [31] and widely referenced, is presented. The model assumes perfect power control in both the uplink and the downlink and does not capture the impact of the soft handoff and macro diversity on cell capacity. For further reading on the effect of soft handoff or imperfect power control, see, for example, [34–35, 37–41, 46–49].

Before presenting the analytical approach, it is important to note that system capacity studies are essential for wireless systems because of the interference induced by users and connections on each other. This is especially important when analyzing the performance of TCP over wireless systems and especially for the UMTS-HSDPA system in this book. Determining the TCP performance of each connection independently or separately as typically done for wired networks and Internet is insufficient for wireless. Such analysis does not capture the interaction or interference between wireless links. Modeling of TCP performance over the HSDPA system in UMTS is addressed in Chapters 4 and 7.

3.14.1 Uplink Capacity

The first step to evaluate the cell capacity is to determine the expression of the SINR at the receiver. On the uplink, the SINR can be approximated by

$$SINR = \frac{W}{R} \frac{p_i}{\sum_{k \neq i}^{N} p_k + I_{inter} + \eta}, \quad (3.3)$$

where η is the thermal noise, p_i is the received power from user i, N is the number of users, R is the bit rate, W is the chip rate (3.84 Mchips/sec), and I_{inter} is the inter cell interference. If the same service is considered for all users (i.e., the same bit rate and same received power p achieved by a perfect power control), the SINR can be simplified as follows:

$$SINR = \frac{W}{R} \frac{1}{N - 1 + I_{inter}/p + \eta/p}. \quad (3.4)$$

It was shown in [31] that I_{inter}/p can be approximated by a gaussian variable with mean value $m = 0.247N$ and variance $v = 0.078N$. Therefore, the probability of outage can be evaluated as follows:

$$P(SINR > \gamma) = P(I_{inter}/p > \delta - N) = Q\left(\frac{\delta - N - 0.247N}{\sqrt{0.078N}}\right), \quad (3.5)$$

where $\delta = \frac{W}{R}\frac{1}{\gamma} - \frac{\eta}{p}$ and γ is the target SINR (4.5 dB for data at 32 Kbps and 7 dB for voice). The outage probability according to the number of cell users for 32 Kbps data service is depicted in Figure 3.15.

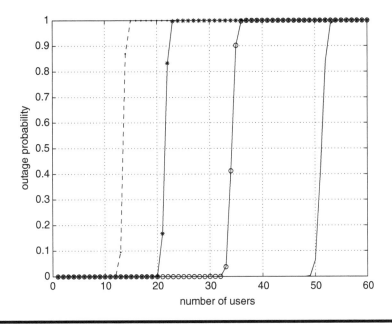

Figure 3.15 Outage probability according to the number of users in the uplink.

3.14.2 Downlink Capacity

In the downlink, the SINR can be approximated using the following expression:

$$SINR = \frac{W}{R_i} \frac{P_i G_{ij}}{\alpha(P_{cell})G_{ij} + \sum_{l\neq j}(P_{cell}G_{il})}, \tag{3.6}$$

where W is the chip rate, R_i is the user bit rate, G_{ij} is the path gain factor including shadowing between user i and cell j, P_i is the transmitted power of user i, P_{cell} is the total cell transmitted power, and $\sum_{l\neq j} P_{cell}G_{il}$ is the other cell interference. Parameter α is the orthogonality loss factor ($\alpha = 0.4$ in macrocell and 0.06 in microcell environment [50]). Thermal noise is considered to be negligible compared to received interference. G_{ij} is given by $d_{ij}^{-\mu} 10^{s_{ij}/10}$, where d_{ij} is the distance between mobile and node B of cell j, μ is the path loss slope, and s_{ij} corresponds to log-normal shadowing with zero mean and standard deviation σ ($\sigma = 10$ dB).

By noting $f_i = \sum_{l\neq j} G_{il}/G_{ij}$, the SINR expression can be simplified as follows:

$$SINR = \frac{W}{R} \frac{P_i}{\alpha P_{cell} + P_{cell} f_i}. \tag{3.7}$$

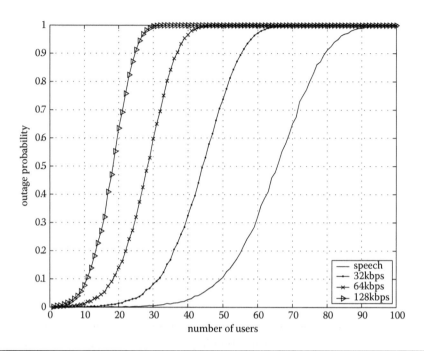

Figure 3.16 Outage probability according to the number of users in the downlink.

An outage occurs if the sum of needed powers exceeds the total cell power ($\sum P_i > P_{cell}$). If γ is the target SINR, the outage probability can then be given by

$$P_{outage} = P\left(\sum_{i=1}^{N} f_i > \frac{W}{R\gamma} - N\alpha\right). \tag{3.8}$$

In the downlink, the number of users in each cell is limited and the gaussian approximation cannot be used. The best way to evaluate equation (3.8) and to obtain more reliable results is to proceed to a Monte Carlo simulation. Note that in [31], an upper bound of this probability is evaluated, but this bound is loose and provides approximate results that are not accurate enough. Monte Carlo simulation results on outage probability as a function of accepted speech users in an arbitrary cell are reported in Figure 3.16.

References

1. Bormann, C., C. Burmeister, H. Fukushima, H. Hannu, L-E. Jonsson, R. Hakenberg, T. Koren, K. Le, Z. Liu, A. Martensson, A. Miyazaki, K. Svanbro, T. Wiebke, T. Yoshimura, and H. Zheng. 2001. RObust Header Compression

(ROHC): Framework and Four Profiles: RTP, UDP, ESP, and Uncompressed RFC 3095. (July).

2. Tanner, R., and J. Woodard. 2004. *WCDMA Requirements and Practical Design*, Chichester, West Sussex, England: The Atrium.

3. Holma, H., and A. Toskala. 2004. *WCDMA for UMTS. Radio Access for Third Generation Mobile Communications*, 3rd ed. Chichester, West Sussex, England.

4. 3GPP TS 23.107 V6.2.0. 2004. Quality of Service (QoS) Concept and Architecture. Release 6, December.

5. Laiho, J., A. Wacker, and T. Novosad. 2002. *Radio Network Planning and Optimisation for UMTS*. London: John Wiley & Sons.

6. ITU-T, Recommendation G114. One-Way Transmission Time. 1996.

7. 3GPP TS 22.105 V6.4.0. 2005. Services and service capabilities. Release 6, September.

8. Ahvonen, K. IP Telephony Signalling in a UMTS all IP Network. 2000. Master's thesis, Helsinki University of Technology, November.

9. 3GPP TS 23.101 V6.0.0. 2004. Universal Mobile Telecommunications System (UMTS) Architecture. Release 6, December.

10. 3GPP TS 23.002 V6.7.0. 2005. Network Architecture. Release 6, March.

11. 3GPP TS 23.121 V3.6.0. 2002. Architectural Requirements for Release 1999. Release 1999, June.

12. 3GPP TS 25.401 V6.5.0. 2004. UTRAN Overall Description. Release 6, December.

13. 3GPP TS 25.410 V6.2.0. 2004. UTRAN Iu Interface: General Aspects and Principles. Release 6, December.

14. 3GPP TS 25.420 V6.3.0. 2005. UTRAN Iur Interface General Aspects and Principles. Release 6, March.

15. 3GPP TS 25.430 V6.5.0. 2005. UTRAN Iub Interface: General Aspects and Principles. Release 6, June.

16. 3GPP TS 25.301 V6.2.0. 2005. Radio Interface Protocol Architecture. Release 6, March.

17. 3GPP TS 25.321 V6.5.0. 2005. Medium Access Control (MAC) Protocol Specification. Release 6, June.

18. 3GPP TS 25.211 V6.6.0. 2005. Physical Channels and Mapping of Transport Channels onto Physical Channels (FDD). Release 6, September.

19. 3GPP TS 25.302 V6.3.0. 2005. Services Provided by the Physical Layer. Release 6, March.

20. 3GPP TS 25.201 V6.2.0. 2005. Physical Layer—General Description. Release 6, June.

21. 3GPP TS 25.322 V6.3.0. 2005. Radio Link Control (RLC) Protocol Specification. Release 6, March.

22. 3GPP TS 25.331 V6.7.0. 2005. Radio Resource Control (RRC) Protocol Specification. Release 6, September.

23. 3GPP TS 25.323 V6.3.0. 2005. Packet Data Convergence Protocol (PDCP) specification. Release 6, September.

24. Degermark, M., B. Nordgren, S. Pink. 1999. IP Header Compression. RFC 2507, February.

25. 3GPP TS 25.324 V6.4.0. 2005. Broadcast/Multicast Control (BMC). Release 6, September.

26. 3GPP TS 25.346 V6.6.0. 2005. Introduction of the Multimedia Broadcast Multicast Service (MBMS) in the Radio Access Network (RAN). Release 6, September.

27. Reed, Irving S., and Xuemin Chen. 1999. *Error-Control Coding for Data Networks*. Boston: Kluwer Academic.

28. Bertsekas, D., and R. Gallager. 1992. *Data Networks*, 2d ed. Englewood Cliffs, NJ: Prentice Hall.

29. 3GPP TS 25.214 V6.5.0. 2005. Physical Layer Procedures (FDD). Release 6, March.

30. 3GPP TR 25.922 V6.0.1. 2004. Radio Resource Management Strategies. Release 6, April.

31. Gilhousen, S. Klein, I. M. Jacobs, R. Padovani, A. J. Viterbi, L. A. Weaver, and C. E. Wheatley. 1991. On the Capacity of CDMA System. *IEEE Transactions on Vehicular Technology* 40, no. 2: 303–12 (May).

32. Viterbi, A. M., and A. J. Viterbi. 1993. Erlang Capacity of a Power Controlled CDMA System. *IEEE Journal on Selected Areas in Communications* 11, no. 6: 892–900 (August).

33. Viterbi, A. J., A. M. Viterbi, and E. Zehavi. 1994. Other-Cell Interference in Cellular Power-Controlled CDMA. *IEEE Transactions on Communications* 42, no. 234: 1501–4 (Part 3).

34. Lee, C., and R. Steele. 1998. Effect of Soft and Softer Handoffs on CDMA System Capacity. *IEEE Transactions on Vehicular Technology* 47, no. 3: 830–41 (August).

35. Leibnitz, K. 2003. Analytical Modeling of Power Control and Its Impact on Wideband CDMA Capacity and Planning. PhD diss., University of Wurzburg, February.

36. Amaldi, E., A. Capone, and F. Malucelli. 2003. Planning UMTS Base Station Location: Optimization Models with Power Control and Algorithms. *IEEE Transactions on Wireless Communications* 2, no. 5:939–952 (September).

37. Laiho, J., A. Wacker, and T. Novosad, eds. 2001. *Radio Network Planning and Optimization for UMTS*. London: John Wiley and Sons.

38. Momentum Project. 2001. Models and Simulations for Network Planning and Control of UMTS. http://momentum.zib.de.

39. Akhtar, S., and D. Zeghlache. 1999. Capacity Evaluation of the UTRA WCDMA Interface. *IEEE Vehicular Technology Conference* 2:914–8 (September 19–22).

40. Malik, S. A., and D. Zeghlache. 2002. Resource Allocation for Multimedia Services on the UMTS Downlink. *IEEE International Conference on Communications* 5:3076–80 (April 28–May 2).

41. Malik, S. A., and D. Zeghlache. 2002. Downlink Capacity and Performance Issues in Mixed Services UMTS WCDMA Networks Malik. *IEEE Vehicular Technology Conference* 4:1824–8 (May 6–9).

42. Lee, W. C. Y. 1991. Overview of Cellular CDMA. *IEEE Transactions on Vehicular Technology* 40, no. 2:291–302 (May).

43. Sampath, A., N. B. Mandayam, and J. M. Holtzman. 1997. Erlang Capacity of a Power Controlled Integrated Voice and Data CDMA System. *IEEE Vehicular Technology Conference* 3:1557–61 (4–7 May 1997).
44. Lee, Seung Joon, and Dan Keun Sang. 1998. Capacity Evaluation for DS-CDMA Systems with Multiclass On/OFF Traffic. *IEEE Communication Letters* 2, no. 6:153–5.
45. Ayyagari, D., and A. Ephramides. 1999. Cellular Multicode Capacity for Integrated (Voice and Data) Services. *IEEE Journal on Selected Areas in Communications* 17, no. 5:928–38 (May).
46. Viterbi, A. J., A. M. Viterbi, and E. Zehavi. 1994. Soft Handoff Extends CDMA Cell Coverage and Increases Reverse Link Capacity Viterbi. *IEEE Journal on Selected Areas in Communications* 12, no. 8:1281–8 (October).
47. Zhang, N., and J. M. Holtzman. 1998. Analyis of a CDMA Soft-Handoff Algorithm. *IEEE Transactions on Vehicular Technology* 47, no. 2:710–15 (May).
48. Zou, J., and V. K. Barghava. 1995. On Soft-Handoff, Erlang Capacity, and Service Quality of a CDMA Cellular System. In Proc. of the 7th IEEE International Symposium on Personal, Indoor and Mobile Radio Communications (IEEE PIMRC), 603–7.
49. Rege, K., S. Nanda, C. Weaver, and W. Peng. 1996. Fade Margins for Soft and Hard Handoffs. *Wireless Networks* 2:277–88.
50. 3GPP TR 25.942 V6.4.0. 2005. Radio Frequency (RF) System Scenarios. Release 6, March.

Chapter 4

High-Speed Downlink Packet Access

The UMTS system, presented in Chapter 3, proposed for third-generation cellular networks in Europe is meant to provide enhanced spectral efficiency and data rates over the air interface. The objective for UMTS, known as WCDMA in Europe and Japan, is to support data rates up to 2 Mbps in indoor or small-cell outdoor environments and up to 384 Kbps in wide-area coverage for both packet and circuit-switched data. The 3GPP, responsible for standardizing the UMTS system, realized early on that the first releases for UMTS would be unable to fulfill this objective. This was evidenced by the limited achievable bit rates and aggregate cell capacity in Release 99. The original agenda and schedule for UMTS evolution has been modified to meet these goals by gradual introduction of advanced radio, access, and core network technologies through multiple releases of the standard. This phased roll-out of UMTS networks and services also would ease the transition from second- to third-generation cellular for manufacturers, network, and service providers. In addition, to meet the rapidly growing needs in wireless Internet applications, studies initiated by 3GPP since 2000 not only anticipated this needed evolution but also focused on enhancements of the WCDMA air interface beyond the perceived third-generation requirements.

The high-speed downlink packet access (HSDPA) system [1–5] has been proposed as one of the possible long-term enhancements of the UMTS standard for downlink transmission. It has been adopted by the 3GPP and will be used in Europe starting in 2006 and 2007. HSDPA introduces first-adaptive modulation and coding, retransmission mechanisms over the radio link and fast packet scheduling, and, later on, multiple transmit-and-receive antennas. This chapter describes the HSDPA system and some of these related advanced radio techniques. Similar enhancements are envisaged for the UMTS uplink but are not covered here. Readers can refer to the 3GPP standard for UMTS and especially the high-speed uplink packet access (HSUPA) to find out more about data rate and capacity enhancements for this link.

4.1 HSDPA Concept

Chapter 2 illustrated the importance of the interference control and management in a CDMA-based system. This can be performed at the link level by some enhanced receiver structures called MUD, which is used to minimize the level of interference at the receiver. At the network level, good management of the interference can be provided by an enhanced power control and associated call admission control (CAC) algorithms.

Note that the MUD technique is essentially used on the uplink in actual CDMA-based systems (UMTS, IS95,...) since the node B has knowledge of the CDMA codes used by all users. In the downlink, informing each user about the spreading codes used by other users would increase system complexity. Detection complexity increases with the number of users in the system. This has direct impact on the terminal energy consumption (battery lifetime) and on the response time of the detector (detection delay). Consequently, in UMTS, the MUD technique is used only in the uplink. In the downlink, interference is managed at the network level using fast power control and CAC.

This philosophy of simultaneously managing the interference at the network level for dedicated channels leads to limited system efficiency. Fast power control used to manage the interference increases the transmission power during the received signal fades. This causes peaks in the transmission power and subsequent power rises that reduce the total network capacity. Power control imposes provision of a certain headroom, or margin, in the total node B transmission power to accommodate variations [6]. Consequently, system capacity remains insufficient and unable to respond to the growing need in bit rates due to the emergence of Internet applications. A number of performance-enhancing technologies must be included in the UMTS standard to achieve higher aggregate bit rates in the downlink and to increase the spectral efficiency of the entire system. These techniques include AMC, fast-link adaptation, hybrid ARQ, and fast scheduling. MUD and MIMO antenna solutions can also be included—which is expected in later releases of UMTS—to further improve system performance and efficiency.

The use of higher-order modulation and coding increases the bit rate of each user but requires more energy to maintain decoding performance at the receiver. Hence, the introduction of fast-link adaptation is essential to extract any benefit from introducing higher-order modulation and coding in the system. The standard link adaptation used in current wireless system is power control. However, to avoid power rise as well as cell transmission power headroom requirements, other link adaptation mechanisms to adapt the transmitted signal parameters to the continuously varying channel conditions must be included. One approach is to tightly couple AMC and scheduling. Link adaptation to radio channel conditions is the baseline philosophy in HSDPA, which serves users having favorable channel conditions. Users with bad channel conditions should wait for improved conditions to be served. HSDPA adapts in parallel the modulation and the coding rates according to the instantaneous channel quality experienced by each user.

AMC still results in errors due to channel variations during packet transmission and in feedback delays in receiving channel quality measurements. A hybrid ARQ scheme can be used to recover from link adaptation errors. With HARQ, erroneous transmissions of the same information block can be combined with subsequent retransmission before decoding. By combining the minimum number of packets needed to overcome the channel conditions, the receiver minimizes the delay required to decode a given packet. There are three main schemes for implementing HARQ: chase combining, in which retransmissions are a simple repeat of the entire coded packet; incremental redundancy IR, in which additional redundant information is incrementally transmitted; and self-decodable IR, in which additional information is incrementally transmitted but each transmission or retransmission is self-decodable.

The link-adaptation concept adopted in HSDPA implies the use of time-shared channels. Therefore, scheduling techniques are needed to optimize the channel allocation to the users. Scheduling is a key feature in the HSDPA concept and is tightly coupled to fast-link adaptation. Note that the time-shared nature of the channel used in HSDPA provides significant trunking benefits over DCH for bursty high data-rate traffic.

The HSDPA shared channel does not support soft handover due to the complexity of synchronizing the transmission from various cells. Fast cell selection can be used in this case to replace the soft handover. It could be advantageous to be able to rapidly select the cell with the best SIR for the downlink transmission.

MIMO antenna techniques could also provide higher spectral efficiency. Such techniques exploit spatial or polarization decorrelations over multiple channels to achieve fading diversity gain. There are two types of MIMO: blast MIMO and space–time coding.

HSDPA can be seen as a mixture of enhancement techniques tightly coupled and applied on a combined CDMA-TDMA (Time Division Multiple

Access) channel shared by users [6]. This channel, called HS-DSCH, is divided into slots called transmit time intervals (TTIs), each one equal to 2 ms. The signal transmitted during each TTI uses the CDMA technique. Since link adaptation is used, "the variable spreading factor is deactivated because its long-term adjustment to the average propagation conditions is not required anymore" [6]. Therefore, the spreading factor is fixed and equal to 16. The use of relatively low spreading factor addresses the provision for increased applications bit rates.

Finally, the transmission of multiple spreading codes is also used in the link-adaptation process. However, a limited number of Wash codes is used due to the low spreading adopted in the system. Since all these codes are allocated in general to the same user, MUD can be used at the user equipment to reduce the interference between spreading codes and to increase the achieved data rate. This is in contrast to traditional CDMA systems where MUD techniques are used in the uplink only.

4.2 HSDPA Structure

As illustrated in Section 4.1, HSDPA relies on a new transport channel, called HS-DSCH, which is shared between users. Fast link adaptation combined with time domain scheduling takes advantage of the short-term variations in the signal power received at the mobile, so that each user is served on favorable fading conditions. The TTI value, fixed at 2 ms in the 3GPP standard, allows this fast adaptation to the short-term channel variations. To avoid the delay and the complexity generated by the control of this adaptation at the RNC, the HS-DSCH transport channel is terminated at the node B, unlike the transport channels in UMTS, which are terminated at the RNC. To handle the fast link adaptation combined with scheduling and HARQ at the node B, a new MAC entity, called MAC-hs (high speed), has been introduced at the node B.

The general architecture of the radio protocol is depicted in Figure 4.1 [1]. The MAC-hs is located below the MAC-c/sh entity in the controlling RNC. The MAC-c/sh provides functions to HSDPA that already exist in UMTS. MAC-d is still included in the serving RNC. The HS-DSCH frame protocol (HS-DSCH FP) handles the data transport from S-RNC to C-RNC, and between C-RNC and node B. Another alternative configuration, presented in Figure 4.2 [1], is also proposed in the 3GPP standards where the S-RNC is directly connected to the node B—that is, without any user plane of C-RNC. Note that in both configurations, the HSDPA architecture does not affect protocol layers above the MAC layer.

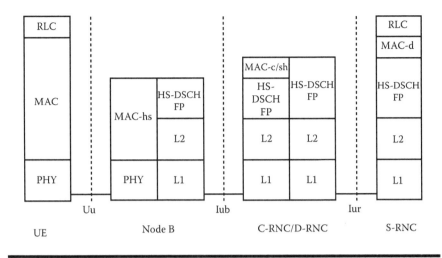

Figure 4.1   Radio interface protocol architecture of HSDPA system.

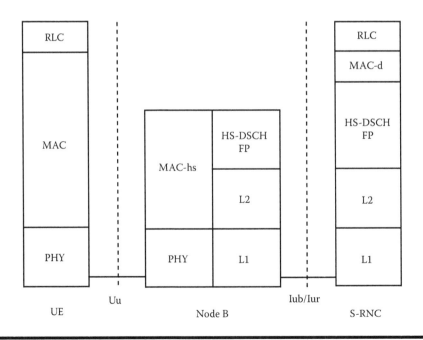

Figure 4.2   Radio interface protocol architecture of HSDPA system.

4.3 Channels Structure

HSDPA consists of a time-shared channel between users and is consequently suitable for bursty data traffic. HSDPA is basically conceived for nonreal-time data traffic. Research is actually ongoing to handle streaming traffic over HSDPA using improved scheduling techniques.

In addition to the shared data channel, two associated channels called HS-SCCH and HS-DPCCH are used in the downlink and the uplink to transmit signalling information to and from the user. These three channels in HSDPA—HS-DSCH, HS-SCCH, and HS-DPCCH—are now described in more detail.

4.3.1 HS-DSCH Channel

The fast adaptation to the short-term channel variations requires handling of fast link adaptation at the node B. Therefore, the data transport channel HS-DSCH is terminated at the node B. This channel is mapped onto a pool of physical channels called HS-PDSCH to be shared among all the HSDPA users on a time and code multiplexed manner [7,8] (see Figure 4.3). Each physical channel uses one channelization code of fixed spreading factor equal to 16 from the set of 15 spreading codes reserved for HS-DSCH transmission. Multicode transmission is allowed, which translates to a mobile user being assigned multiple codes in the same TTI depending on the user equipment capability. Moreover, the scheduler may apply code multiplexing by transmitting separate HS-PDSCHs to different users in the same TTI.

The transport-channel coding structure is reproduced as follows. One transport block is allocated per TTI, so that no transport block concatenation—such as in UMTS DCH-based transmission—is used. The size of transport block changes according to the modulation and coding scheme (MCS) selected using the AMC technique. To each transport block, a cyclic redundancy check (CRC) sequence with 24 bits is added.

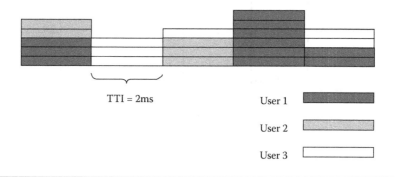

TTI = 2ms   User 1
               User 2
               User 3

Figure 4.3   Code multiplexing example over HS-DSCH.

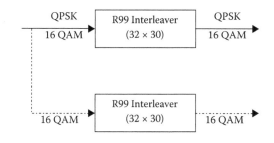

Figure 4.4   Interleaver structure for HSDPA.

Since errors occur in bursts, one CRC sequence per transport block (i.e., per TTI) is sufficient. Once the CRC sequence is attached, the transport block bits are bit scrambled and are segmented into blocks to apply turbo block encoding. The code-block size depends on the turbo coding rate and can reach a maximum value of 5114 bits [9]. The coding rate changes according to the MCS scheme selected by the link-adaptation technique. At the turbo encoder output, rate matching is applied by the physical layer HARQ functionality. After matching the number of bits to the number of bits in the allocated HS-DSCH physical channels, segmentation divides the resulting bits among the HS-DSCH physical channels. The bit sequence obtained for each physical channel is then interleaved using one step interleaver with fixed size 32×30 (i.e., 32 rows and 30 columns) (see Figure 4.4 [9]). Finally, the resulting bit sequence is modulated using 16-ary quadrature amplitude modulation (16QAM) or quadrature phase shift keying (QPSK) according to the MCS scheme selected [10,11].

4.3.2   HS-SCCH Channel

The downlink signaling related to the HS-DSCH is transmitted over the HS-SCCH. The signaling information carried by the HS-SCCH contains essentially the transport format resource indicator (TFRI) and the HARQ information of the HS-DSCH channel. The TFRI includes the channelization codes used by the HS-DSCH, the modulation scheme, and the transport block size. The HARQ information consists of the HARQ new data indicator, the HARQ process identifier, and the redundancy and constellation version. Since the HS-DSCH channel is shared among users, the user equipment identity is sent over the HS-SCCH to indicate the identity of the user for which the HS-DSCH is allocated during the TTI. Note that the user equipment identity is given by the 16-bit HS-DSCH radio network temporary identifier (H-RNTI) defined by the RRC [12,13].

The information carried by this channel is split into two parts [9]:

■ Part I contains (1) channelization codes (7 bits); and (2) modulation scheme (1 bit).

- Part II includes (1) transport block size (6 bits); (2) HARQ process identifier (3 bits); (3) HARQ new data indicator (1 bit); and (4) redundancy and constellation version (3 bits). The HS-SCCH also contains a CRC attachment, which consists of 16 bits calculated over part I and II and appended to part II, as in Figure 4.5 [9], where
  - X_{ccs}: 7 bits of channelization code set information
  - X_{ms}: 1 bit of modulation scheme
  - X_{tbs}: 6 bits of transport block size information
  - X_{hp}: 3 bits of HARQ process information
  - X_{rv}: 3 bits of redundancy version information
  - X_{nd}: 1 bit of new data indicator
  - X_1: 8 bits of input to rate 1/3 convolutional encoder
  - Z_1: 48 bits of output
  - R_1: 40 bits after puncturing

Part I and II are encoded separately using Release 99 convolutional code with coding rate equal to 1/3 and 8 tail bits. After convolutional coding, interleaving and rate matching to 120 HS-SCCH channel bits (three slots) is applied. The interleaving and rate matching is carried out separately for the two parts of the coded HS-SCCH to allow for early extraction of the time-critical information of Part I of the HS-SCCH information [9]. Note that the postconvolution of part I is scrambled by a code generated from the user equipment identity, which is encoded using Release 99 convolutional code with coding rate equal to 1/2 and 8 tail bits. Of the resulting 48 bits, 8 are punctured using the same rate matching rule as for part I of the HS-SCCH. The resulting part I (40-bit sequence) is mapped into the first slot of the HS-SCCH TTI. The resulting part II is carried over the second and the third slots. Note that the HS-SCCH is spread with a fixed spreading factor of 128 so that the bit rate over this channel is fixed at 60 Kbps (i.e., 120 bits per TTI), as in Figure 4.6 [8].

Finally, this channel can be transmitted at a fixed power or can use power control. The decision to use power control is entirely left to the implementation.

4.3.3 HS-DPCCH Channel

In the uplink, signaling information has to be transmitted for the HARQ acknowledgment and the feedback measurement. The use of fast link adaptation on the HS-DSCH channel requires knowledge of the channel quality during the transmission. The user equipment measures the channel quality on the CPICH and sends the result to the node B. This procedure is explained in Section 4.5. The use of HARQ requires an acknowledgment message from the user to the node B so that the node B retransmits the erroneously received packet or a new packet.

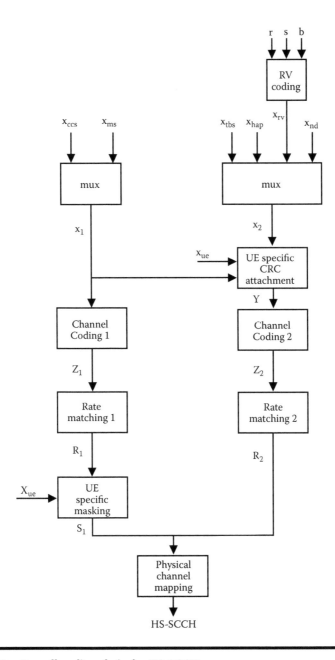

Figure 4.5 Overall coding chain for HS-SCCH.

This signaling information is carried by the HS-DSCH associated uplink dedicated control channel (HS-DPCCH). This channel is spread, with a spreading factor of 256 (i.e., 30 bits per TTI), and is code multiplexed with the existing dedicated uplink physical channels (DPCH). The HARQ

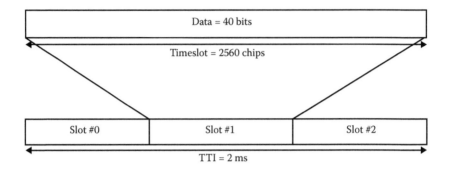

Figure 4.6   Frame structure of HS-SCCH channel.

acknowledgment is a 1-bit acknowledgment/nonacknowledgement (ACK/NACK) indication repeated 10 times and transmitted in one slot. The HARQ acknowledgement field is gated off when there is no ACK/NACK information being sent. The measurement feedback information contains a channel quality indicator (CQI) that may be used to select transport format and resource by HS-DSCH serving the node B, according to CQI tables specified in [11]. It is essential information needed for fast link adaptation and scheduling. The channel quality information, consisting of 5 bits, is coded using a (20,5) code transmitted over two slots (see Figure 4.7 [8]).

4.3.4   Timing of HSDPA Channels

The timing relation between the HSDPA channels is presented in Figure 4.8 [8]. In the downlink, the HS-SCCH is received 2 slots (1.334 ms) before the HS-DSCH channel. Therefore, the user gets the time to decode the first part

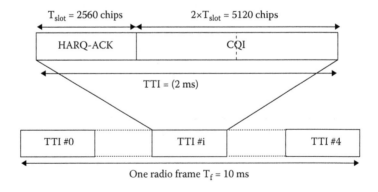

Figure 4.7   Frame structure of HS-DPCCH channel.

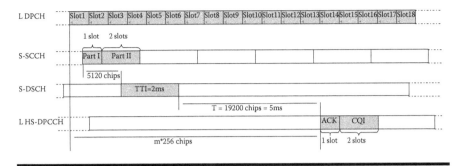

Figure 4.8   Channels transmission timing in HSDPA.

of the HS-SCCH, which essentially contains the channelization codes of the HS-DSCH channel scrambled by the user equipment identity. Once this first part is decoded, the user is able to start decoding the HS-DSCH.

In the uplink, the HS-DPCCH starts m*256 chips after the start of the uplink DPCCH with m selected by the user equipment such that the ACK/NACK transmission (of duration 1 timeslot) commences within the first 0–255 chips after 7.5 slots following the end of the received HS-DSCH. The 7.5 slots represent the user equipment processing time after the reception of the HS-DSCH channel. This time is required due to timing constraints imposed by detection with relatively high complexity such as MUD and HARQ mechanism, or soft combining. The 0–255 chips delay is needed because the HS-DPCCH is code multiplexed with other physical channels transmitted by the same user. HS-DPCCH and the other DPCCH channels use the same scrambling codes, and a physical synchronization between these channels is required to reduce the interference at the receiver (as explained in Chapter 2). The term *physical synchronization* means that the spreading sequences of these channels must start at the same instant. Since the spreading factors of HS-DPCCH and DPCCH are fixed at 256, the synchronization is achieved by introducing a delay of 0–255 chips. Note that the physical synchronization is different from the transport synchronization when the transport channels start their transmissions at the same time.

4.4   MAC-hs

The use of fast adaptation to the short-term channel variations requires handling of the HSDPA transport channels by the node B. Therefore, a medium access control-high speed entity has been introduced in the node B. The MAC-hs also stores the user data to be transmitted across the air interface. This imposes some constraints on the minimum buffering capabilities of the node B. The transfer of the data queues to the node B creates the need for a flow control mechanism (HS-DSCH frame protocol) to handle data transmission from the serving RNC to the controlling RNC, if the Iur interface is involved, and between the controlling RNC and the node B. The overall MAC architecture at the UMTS, UTRAN side and the user equipment side is presented, respectively, in Figures 4.9 and 4.10 [1,14].

4.4.1 MAC Architecture at the UTRAN Side

At the UTRAN side, the data to be transmitted on the HS-DSCH channel is transferred from MAC-c/sh to the MAC-hs via the Iub interface in case of configuration with MAC-c/sh, or from the MAC-d via Iur/Iub in case of configuration without MAC-c/sh. As specified by the 3GPP [1,14], the MAC-hs entity is in charge of four logical functionalities:

1. Flow control: The presence of data queues in the node B creates the need for a flow control mechanism (HS-DSCH frame protocol) to handle data transport between MAC-c/sh and MAC-hs (configuration with MAC-c/sh) or MAC-d and MAC-hs (configuration without MAC-c/sh) taking the transmission capabilities of the air interface into account in a dynamic manner. The design of such flow control is a nontrivial task, since such functionality, in cooperation with the packet scheduler, is intended to ultimately regulate the users' perceived service to fulfill the QoS allocated to the user depending on service class.

2. Scheduling/priority handling: This function manages HS-DSCH channel allocation between users as well as between HARQ entities and data flows of the same user according to services priority and wireless channel conditions. Based on status reports, from associated uplink signaling, the new transmission or retransmission of a given user is determined. To maintain proper transmission priority, a new transmission can be initiated on a HARQ process at any time. However, it is not permitted to schedule new transmissions, including retransmissions originating in the RLC layer, within the same TTI, along with retransmissions originating from the HARQ layer.

3. HARQ: This entity handles the HARQ functionality for users. In other words, it supports multiple HARQ processes of stop-and-wait HARQ protocols. One HARQ entity per user is needed and one HARQ process is active during each TTI. The HARQ technique is explained in detail in Section 4.6.

4. TFRI selection: This entity selects the appropriate transport format and resource combination (i.e., the modulation and coding schemes MCS, described in Section 4.5) for the data to be transmitted on HS-DSCH. The selected TFRI selected changes from TTI to TTI according to wireless channel conditions.

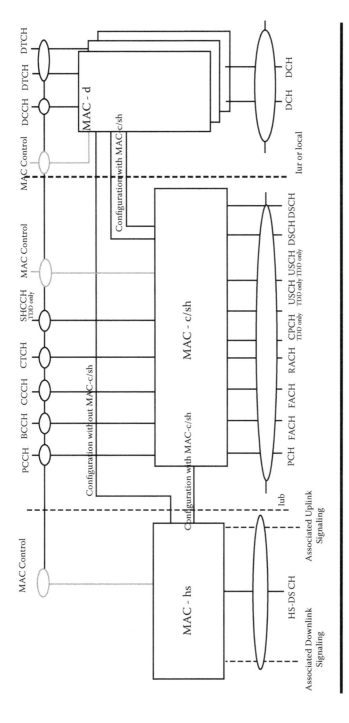

Figure 4.9 UTRAN side MAC architecture with HSDPA.

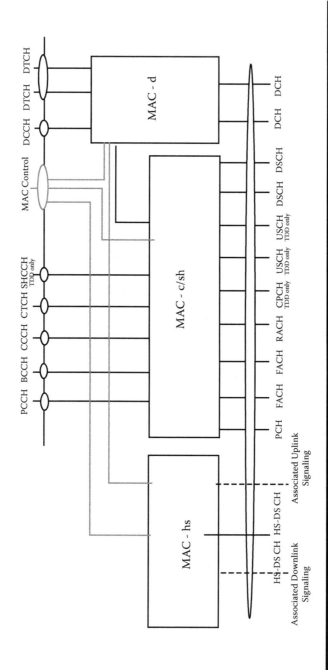

Figure 4.10   UE side MAC architecture with HSDPA.

4.4.2 MAC Architecture at the User Equipment Side

At the user equipment side, the MAC-d entity is modified with the addition of a link to the MAC-hs entity. The links to MAC-hs and MAC-c/sh cannot be configured simultaneously in one user equipment. The mapping between MAC-d and the reordering buffer in MAC-hs is configured by higher layers. The two essential entities of the MAC-hs are [1,14]

1. The HARQ entity, which is responsible for handling all the tasks of the HARQ protocol such as generating ACKs and NACKs messages. HSDPA uses "N-Channel Stop and Wait HARQ" protocol. This results in N HARQ processes to be handled by the HARQ peer entities in the user equipment and the UTRAN. Only one HARQ process is active per HS-DSCH per TTI. The configuration of the hybrid ARQ protocol is provided by RRC over the MAC-control service access point (SAP) [12].

2. The reordering entity, which organizes received data blocks and delivers them in sequence to higher layers on reception. One reordering entity per priority class can be used as specified by the 3GPP. The use of this reordering entity may increase the overall delay of the data reception at the upper layers. However, this entity has an important impact on the reduction of the TCP performance degradation over wireless links. Out-of-order received data at the transport layer are misinterpreted by TCP as triple duplicate congestion. This erroneous interpretation results in retransmissions and reduction of congestion window size, in other words in throughput reduction (see Chapters 5 to 7). In this MAC-hs reordering entity, out-of-sequence data blocks can be delivered to the upper layers when the delay of receiving the appropriate block (i.e., having the adequate sequence number) exceeds a certain limit. This limit is determined by a timer configured by RRC. When the timer corresponding to a given block (i.e., MAC-hs PDU) expires, the HARQ entity in the user equipment delivers the data blocks in the buffer, including the subsequent received blocks, to higher layers since the missing blocks will not be retransmitted by the HARQ entity in the UTRAN.

4.5 Fast Link Adaptation

As explained in Chapter 1, the wireless channel in cellular systems is a composite multipath/shadowing channel. The radio waves follow several paths from the transmitter to reach the destination. Fading occurs when many paths interfere, adding up to a composite signal exhibiting short time signal power variations at the receiver. This power could be weaker or stronger than the required power needed to achieve a given user QoS.

The link quality between the transmitter and the receiver is also affected by slow variations of the received signal amplitude mean value due to shadowing from terrains, buildings, and trees.

To deal with the problems caused by multipath fast fading, the existing wireless systems use diversity techniques such as long interleaving, robust channel coding, frequency hopping, and direct-sequence spread spectrum. These techniques are based on one concept: averaging the temporary fading effect over all the transmission time and the bandwidth so that bad conditions are compensated by good conditions. The spread-spectrum technique—frequency hopping and direct sequence—spreads the signal bandwidth over a wider frequency spectrum so that only a part of the spectrum is affected by the fading. Interleaving can be seen as spreading technique over time. By reordering the bits before transmission, the information message is spread out over time. Therefore, bursty errors caused by the fading channel are spread out in time so the decoder receives distributed nonbursty random errors that are easier to detect and to correct. The channel coding includes redundancy in the transmitted signal to increase robustness against errors. Introducing more redundancy increases robustness but decreases effective information bit rate. In current systems, the channel coding rate is fixed to deal with the worst case, and the transmission power is adapted to the channel conditions to achieve the application QoS [15].

Since fading results from the addition of different waves from multiple paths, it can potentially be predicted. Channel parameters (i.e., amplitude, phase, frequency) remain stationary for time windows of the order of half a wavelength. For example, at a carrier frequency of 2 GHz in case of UMTS and at a mobile speed of 36 Km/h, the fading pattern could be predicted for a time window of approximately 7.5 ms. Estimation of the channel is feasible over a few ms, and power consequently can be adjusted over such timescales.

In UMTS Release 99, channel estimation is used to adapt the transmission power of each user and every slot (corresponding to a rate of 1500 Hz) to the short-term channel variations. This is achieved at the cost of some power rise and higher interference, as previously explained in Section 4.2.

In HSDPA, the idea is to avoid power adaptation and, hence, power control by approaching the radio resource allocation and sharing from a different angle. Why continue using averaging techniques such as long interleaving and a fixed channel-coding rate to counteract the fast fading if these techniques require high-performance power control? Instead, one can use AMC tightly coupled with fast scheduling so that modulation orders and coding rates are adapted according to estimated channel fading. In addition, the HS-DSCH channel can be allocated to the user with favorable channel conditions. To this avail, a channel quality indicator (CQI) has been introduced in HSDPA (for details, see next section) to enable such intelligent allocation of resources to users.

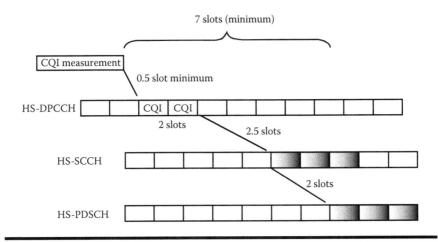

Figure 4.11 Timing relation between CQI measurement and HS-DSCH transmission.

The idea is to measure the channel quality over the CPICH and to transmit the measurement report over the HS-DPCCH channel to the node B, so that scheduling and AMC can act according to the CQI and, hence, can optimize channel resource allocation. The time window between the channel conditions measurement and the resource allocation should not exceed half a wavelength, as indicated previously. In the 3GPP specifications [11], 30 CQIs have been standardized, as described in the next section. The timing relation between the channel measurement over the CPICH and the resource allocation over the HS-DSCH is depicted in Figure 4.11 [16]. The time window between the channel measurement and the start of the transmission over the HS-DSCH channel is seven timeslots. By considering the transmission time of the HS-DSCH channel (one TTI), the overall delay between the radio channel measurement and the end of the packet transmission over the HS-DSCH channel is 10 timeslots. Therefore, HSDPA is suitable for urban area when mobile users move at low speed (less than 40 Km/h). For mobile at higher speeds, the environment conditions change rapidly. Fortunately, in this case the mobiles have a low degree of randomness since they move along known paths (traveling by train, driving down a freeway, etc.). Therefore, special solutions can be performed to predict the channel fading pattern in these cases [15].

In the 3GPP Release 6, enhancements were added to the CQI reporting method by introducing tuneable reporting rates through additional CQI reports during periods of downlink activity and fewer reports at other times [4]. These additional CQIs can be initiated on demand of fast Layer-1 signaling. In addition, a certain number of successive CQI values may be averaged with respect to channel quality at the user equipment. The averaged value is reported and used with the instantaneous CQI measured to select the MCS. The motivation for this technology is to improve the selection of MCS so that the delay due to HARQ retransmissions can be reduced. In addition, the uplink signaling overhead may be reduced and this decreases the uplink interference. Finally, the use of the feature requesting extra CQI transmissions by fast Layer-1 signaling improves the performance of the first packets of a packet call.

4.6 Adaptive Modulation and Coding

As explained earlier, the link adaptation in HSDPA is performed by the use of adaptive modulation and coding. According to the channel quality, the modulation and coding are assigned to the user change so that higher peak data rate and average throughput can be provided. A CQI has been introduced to inform the system about the channel conditions. To guarantee a BLER lower than 10 percent on the HS-PDSCH, each CQI is mapped onto a specific modulation and coding scheme corresponding to a given transport format. The selection of the transport format is performed by the MAC-hs located in the node B. Each transport format or MCS is defined by a [11]

1. Modulation format, which can be either QPSK or 16QAM.
2. Turbo encoder rate, which varies between 0.17 and 0.89. The encoding rate depends on the user equipment capabilities (i.e., maximum number of HS-DSCH codes it can handle) and the desired transport block size (TBS). The different code rates are obtained through puncturing of bits in the turbo encoder of rate 1/3.
3. Number of HS-PDSCH codes allocated to the user, which ranges from 1 to the maximum number of codes supported by the user equipment, dependent on its category. Note that for any user equipment category, this number can not exceed 15 codes. In fact, the spreading factor used in HSDPA is fixed at 16. Therefore, 16 branch codes are available, from which at least one code branch is reserved for signalling and control channels, thus leaving a maximum of 15 codes at best to allocate to a given user.

In the 3GPP specifications [11], 12 user equipment categories are defined according to the maximum number of HS-DSCH the user equipment can handle simultaneously. The maximum transport format that can be allocated to the user according to the user equipment capabilities is depicted in Table 4.1. The transport formats and the corresponding CQIs for

Table 4.1   Maximal AMC Transport Format for Different UE Categories

UE Category	Transport Block Size	Number of HS-PDSCH	Modulation
1	7168	5	16-QAM
2	7168	5	16-QAM
3	7168	5	16-QAM
4	7168	5	16-QAM
5	7168	5	16-QAM
6	7168	5	16-QAM
7	14411	10	16-QAM
8	14411	10	16-QAM
9	17237	12	16-QAM
10	25558	15	16-QAM
11	3319	5	QPSK
12	3319	5	QPSK

user equipment category 10 (i.e., maximum number of 15 simultaneous HS-PDSCH codes it can handled) are shown in Table 4.2 [11].

4.7   HARQ

Fast link adaptation provides the flexibility to match the MCS to the short-term channel variations for each user. However, link adaptation is sensitive to measurement errors, delays in the CQI procedure, and unexpected channel variations. Therefore, the use of ARQ, which is insensitive to CQI measurement errors, is indispensable for tolerating higher error rates to save cell resources, to use higher-order MCS, and to increase the user and the average cell throughput.

The ARQ technique used in UMTS is selective repeat. The retransmissions are performed by the RLC layer in the RNC. Introducing ARQ induces delays in receiving error-free information and unfortunately interacts with higher-layer protocols such as the TCP used to handle end-to-end IP packcet communications between end hosts. If these interactions are not addressed and handled properly, drastic degradation in applications flow rates are experienced. These interactions and means to reduce their negative effects on overall system throughput are addressed in Chapters 6 and 7.

In HSDPA, the ARQ protocol is performed by the MAC-hs entity in the node B. HSDPA uses the ARQ protocol combined with the FEC code so that an erroneous packet is not discarded but instead is softly combined with its retransmissions to reduce the average delay in receiving error-free information. In addition, the tight coupling of HARQ and fast link adaptation limits the excessive use of ARQ—that is, the delay—since retransmissions occur if link adaptation fails to cope with the instantaneous channel conditions.

Table 4.2   CQI Mapping for UE Category 10

CQI Value	Transport Block Size	Number of HS-PDSCH	Modulation
1	137	1	QPSK
2	173	1	QPSK
3	233	1	QPSK
4	317	1	QPSK
5	377	1	QPSK
6	461	1	QPSK
7	650	2	QPSK
8	792	2	QPSK
9	931	2	QPSK
10	1262	3	QPSK
11	1483	3	QPSK
12	1742	3	QPSK
13	2279	4	QPSK
14	2583	4	QPSK
15	3319	5	QPSK
16	3565	5	16-QAM
17	4189	5	16-QAM
18	4664	5	16-QAM
19	5287	5	16-QAM
20	5887	5	16-QAM
21	6554	5	16-QAM
22	7168	5	16-QAM
23	9719	7	16-QAM
24	11418	8	16-QAM
25	14411	10	16-QAM
26	17237	12	16-QAM
27	21754	15	16-QAM
28	23370	15	16-QAM
29	24222	15	16-QAM
30	25558	15	16-QAM

4.7.1   HARQ Types

In HSDPA, three types of HARQ have been studied and standardized in the 3GPP specifications [1,14]: chase combining, Incremental Redundancy (IR), and self-decodable IR.

In [17], Chase showed that the sequence resulting from combining two copies of the same sequence presents a lower error rate than the original sequences. Therefore, instead of discarding erroneous packets, the user equipment proceeds to soft combining, called Chase combining, multiple retransmissions of the same packets before decoding. This concept, developed and standardized in the 3GPP specifications of Release 5 [1], reduces the delays compared to the ARQ used in the UMTS, Release 99. This HARQ algorithm does not interact optimally with the AMC and the fast link adaptation since the multiple retransmissions are the same copies of the first one; that is, the same MCS is used even if the channel conditions change. Consequently, enhanced HARQ algorithms have been introduced in the 3GPP specifications [1,14]. These new schemes rely on incremental redundancy.

In HARQ IR, instead of retransmitting the same copy of the erroneous packet redundant information is incrementally added to retransmissions copies. This represents a better protection of the information packets and copes more with channel conditions and AMC. In this type of HARQ, only a fraction of the information sequence is sent in the retransmission packet—according to the degree of redundancy. The retransmitted packet is not self-decodable and should be combined with the first transmission before decoding. To counteract this problem, a self-decodable IR scheme has been studied and developed by the 3GPP [1,14]. To obtain a self-decodable scheme, incremental redundant information is added to the first sequence, and incremental puncturing is also used so that the receiver can reconstruct the information sequence and can decode each retransmission before soft combining the retransmissions in case of unsuccessful decoding.

Note that in cases of retransmissions, the node B can select a combination for which no mapping exists between the original TBS and the selected combination of MCS. When such cases occur, the TBS index (TBSi) value signalled to the user equipment shall be set to TBSi = 63. TBS is the transport block size value, according to Table 4.2, corresponding to the MCS signaled on the HS-SCCH. Let S be the sum of the two values, S = TBSi + TBS. The transport block size L(S) can be obtained by accessing the position S in Table 4.3 [14].

4.7.2 HARQ Protocol

In UMTS Release 99, selective repeat ARQ is used so that only the erroneous blocks are retransmitted. To achieve selective retransmission, a sequence number is required to identify each block. This requirement increases the system and receiver complexity. By combining the SR protocol and the HARQ techniques in HSDPA, more requirements and complexity are introduced compared to the SR-ARQ in UMTS Release 99. These requirements can be summarized by the following points [18,19]. First, HARQ requires that the receiver must know the sequence number of each block before combining separate retransmissions. Therefore, the sequence number must be encoded separately from the data. In addition, it should be highly protected to cope with the worst channel conditions. Consequently, more signaling bandwidth is consumed. Second, the MAC-hs entity, in charge of handling the HARQ, should deliver data to the upper layers in sequence. Therefore, this entity should store the multiple retransmissions of an erroneous block and the correctly received data that cannot be delivered to

Table 4.3 Transport Block Size for HSDPA

Index	TBSᵃ	Index	TBS	Index	TBS	Index	TBS	Index	TBS	Index	TBS	Index	TBS	Index	TBS
1	137	33	521	65	947	97	1681	129	2981	161	5287	193	9377	225	16630
2	149	34	533	66	964	98	1711	130	3035	162	5382	194	9546	226	16931
3	161	35	545	67	982	99	1742	131	3090	163	5480	195	9719	227	17237
4	173	36	557	68	1000	100	1773	132	3145	164	5579	196	9894	228	17548
5	185	37	569	69	1018	101	1805	133	3202	165	5680	197	10073	229	17865
6	197	38	581	70	1036	102	1838	134	3260	166	5782	198	10255	230	18188
7	209	39	593	71	1055	103	1871	135	3319	167	5887	199	10440	231	18517
8	221	40	605	72	1074	104	1905	136	3379	168	5993	200	10629	232	18851
9	233	41	616	73	1093	105	1939	137	3440	169	6101	201	10821	233	19192
10	245	42	627	74	1113	106	1974	138	3502	170	6211	202	11017	234	19538
11	257	43	639	75	1133	107	2010	139	3565	171	6324	203	11216	235	19891
12	269	44	650	76	1154	108	2046	140	3630	172	6438	204	11418	236	20251
13	281	45	662	77	1175	109	2083	141	3695	173	6554	205	11625	237	20617
14	293	46	674	78	1196	110	2121	142	3762	174	6673	206	11835	238	20989
15	305	47	686	79	1217	111	2159	143	3830	175	6793	207	12048	239	21368
16	317	48	699	80	1239	112	2198	144	3899	176	6916	208	12266	240	21574
17	329	49	711	81	1262	113	2238	145	3970	177	7041	209	12488	241	22147
18	341	50	724	82	1285	114	2279	146	4042	178	7168	210	12713	242	22548
19	353	51	737	83	1308	115	2320	147	4115	179	7298	211	12943	243	22955
20	365	52	751	84	1331	116	2362	148	4189	180	7430	212	13177	244	23370

21	377	53	764	85	1356	117	2404	149	4265	181	7564	213	13415	245	23792
22	389	54	778	86	1380	118	2448	150	4342	182	7700	214	13657	246	24222
23	401	55	792	87	1405	119	2492	151	4420	183	7840	215	13904	247	24659
24	413	56	806	88	1430	120	2537	152	4500	184	7981	216	14155	248	25105
25	425	57	821	89	1456	121	2583	153	4581	185	8125	217	14411	249	25558
26	437	58	836	90	1483	122	2630	154	4664	186	8272	218	14671	250	26020
27	449	59	851	91	1509	123	2677	155	4748	187	8422	219	14936	251	26490
28	461	60	866	92	1537	124	2726	156	4834	188	8574	220	15206	252	26969
29	473	61	882	93	1564	125	2775	157	4921	189	8729	221	15481	253	27456
30	485	62	898	94	1593	126	2825	158	5010	190	8886	222	15761	254	27952
31	497	63	914	95	1621	127	2876	159	5101	191	9047	223	16045		
32	509	64	931	96	1651	128	2928	160	5193	192	9210	224	16335		

[a] TBS: transport block size.

upper layers due to prior erroneous blocks. Consequently, more memory is required in the user equipment.

The increase in complexity and in requirements of the SR-HARQ leads to the adoption of simpler HARQ strategies. The "stop and wait" SW-ARQ protocol is quite simple to implement but since it is inefficient (see Chapter 3), a trade-off between the simple SW and the SR, called N-channel SW, has been developed and standardized for HSDPA [1,14].

The N-channel SW consists of activating N-HARQ processes in parallel, each one using the SW protocol. Hereby, one HARQ process can transmit data on the HS-DSCH while other instances wait for the acknowledgment on the uplink. Using this strategy, the retransmission process behaves as if the SR-HARQ were employed. The advantage of the N-channel SW strategy with respect to the SR protocol is that a persistent failure in a packet transmission affects only one channel, allowing data to be transmitted on the other channels. In addition, compared to the simple SW, the N-channel SW provides the MAC-hs entity with the flexibility to allocate the HS-DSCH channel to the same user if radio conditions are favorable. However, this HARQ strategy imposes timing constraints on the maximum acceptable retransmission delay. The transmitter must be able to retransmit the erroneous packet (N-1)TTIs after the previous transmission. An example of N-channel SW HARQ protocol is presented in Figure 4.12, where the HS-DSCH channel is shared by two users having, respectively, four-channel and one-channel SW-HARQ entities [20].

Increasing the number of processes to N relaxes the timing constraints but simultaneously increases the memory required to buffer the soft samples of each partially received block [13]. Therefore, the number of HARQ instances must be limited. This limit is configured by the upper layers in HSDPA to eight parallel and simultaneous HARQ channels according to the 3GPP specifications [1,14].

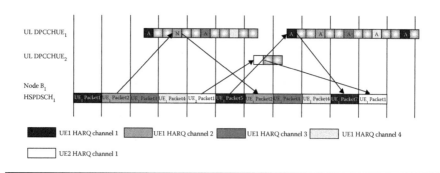

Figure 4.12   Example of N-Channels Stop and Wait HARQ for two active users over HS-DSCH channel.

4.7.3 HARQ Management

The parallel HARQ instances are handled by the MAC-hs entity in the node B. The MAC-hs manages the data-flow control and the reassembly of the RLC PDUs with fixed size into a MAC-hs PDU with variable size. In addition, it performs HARQ and scheduling functionalities.

Once the RLC PDUs are reassembled in a MAC-hs PDU, the resulting PDU is assigned an 8-bit transmission sequence number (TSN) and a HARQ Id, or one of the three available bits. The TSN is carried in the header of the MAC-hs PDU and is transmitted over the HS-DSCH channel; however, the HARQ Id is carried over the HS-SCCH channel. The use of HARQ could generate situations where a packet is decoded error free before one or more previous received packets. Therefore, the TSN is used in the reordering entity, in the MAC-hs layer, to deliver data in sequence to the upper layer. Note that the number of MAC-hs PDUs transmitted is limited by the transmitter window size, up to 32 [21].

The reordering entity could result in a stall avoidance when a persistent failure occurs in the transmission of one block on a given channel of the N-channel SW. This can interact with higher-layer protocols and can drastically affect the performance of the data traffic carried over HSDPA (e.g., streaming). To control the stall avoidance in the user equipment reordering buffer, a timer T is introduced in the MAC-hs reordering entity and is configured by upper layers (see [14] for details).

The timer T is started when a MAC-hs PDU, with a TSN higher than the expected TSN, is correctly received. One timer is active at a given time. Once the MAC-hs PDU, with the expected TSN, is correctly received before the timer expiration, the timer is stopped. If the timer expires before correctly receiving the expected MAC-hs PDU, all correctly received MAC-hs up to the expected MAC-hs PDU and all correctly received packets with TSN higher than the expected TSN are delivered to the disassembly entity in charge of data delivery to the upper layer [14].

4.8 Packet Scheduling

The shared time structure of the HS-DSCH channel supports the use of time scheduling. Fast link adaptation based on AMC tightly coupled with scheduling provides higher transmission rates and average throughput. By allocating the HS-DSCH channel to the user with favorable channel conditions, higher-order MCS are selected, and higher achievable data rate and average throughput are provided. Introducing the MAC-hs entity in the node B for scheduling and using a low TTI value of 2 ms allow better tracking of the short-term variations of the radio channel. Users with temporary good channel conditions are more easily selected.

With the growing demand on data-application services—especially nonreal-time services such as interactive and background—HSDPA, as any wireless system, should provide the capability of supporting a mixture of services with different quality of service requirements. Even if the interactive and background services are seen as best-effort services with no service guarantees, these users still expect to receive data within a certain period of time. The starvation of these users can have a drastic effect on the performance of higher layers such as the TCP layer. Therefore, a minimum service guarantee should be introduced for these services, and HSDPA should achieve some fairness in sharing resources among users and services. Fairness can be defined as meeting the data rate and the delay constraints of the different applications [5,6].

Consequently, the scheduler has two tasks to accomplish: to increase the average throughput by allocating the HS-DSCH channel to the user with favorable channel conditions; and to achieve fairness between services. These two objectives are in contradiction, and there is a risk in achieving one at the expense of the other. A trade-off between fairness and efficiency (e.g., increasing the cell throughput) should be performed by the scheduler.

The scheduling over a time-shared channel has been addressed widely in the literature, and many proposals have been elaborated. The most famous algorithms are Max C/I [2] and proportional fair (PF) proposed in [22,23]. In Max C/I, the node B tracks the channel quality of each user by measuring the SIR on the CPICH channel and allocates the HS-DSCH channel to the user with the best SIR. This algorithm maximizes the cell capacity but presents a problem of fairness between users, especially for users at the cell border. In this context, work in [24] presents a performance comparison of several scheduling algorithms concluding that algorithms providing the highest average cell throughput tend to present the largest variations in throughput per user. To alleviate this problem, PF scheduling has been proposed. PF, which realizes a reasonable trade-off between efficiency and fairness, consists of transmitting to the user with the highest data rate relative to its currently achieved mean data rate. This scheduler, studied in [22–23,25-29], is widely used in currently developed systems. In addition, many other algorithms have been proposed. In [30], six packet-scheduler algorithms were analyzed to assess a trade-off between cell capacity and user fairness. In [31], an opportunistic transmission-scheduling strategy that maximizes the average system throughput given a resource sharing constraint is presented. A study in [32] illustrates a scheduling policy to provide user QoS guarantees in terms of minimum data rates while at the same time achieving some user-diversity gain. Many other proposals, based on the queuing management and the channel state, have been analyzed to achieve fairness. The following list of schedulers is provided for examples: channel state-dependent packet scheduling (CSDPS) [33]; idealized wireless fair queuing algorithm (IWFQ) [34]; channel-condition independent fair queueing (CIF-Q) [35]; server-based fairness algorithm (SBFA) [36]; improved channel state-dependent packet scheduling (I-CSDPS) [37]; channel adaptive fair queueing (CAFQ) [38]; modified largest weighted delay first (MLWDF) [39], code-division generalized processor sharing (CDGPS) [40]; and delay-sensitive dynamic fair queueing (DSDFQ) [41].

4.8.1 Scheduling Constraints and Parameters

In HSDPA, the MAC-hs entity in the node B handles the scheduling functionality using estimated radio channel conditions to enhance cell throughput while satisfying user QoS. Certain constraints concerning the radio channel conditions, the physical available resources in the cell, and user QoS have to be considered in the scheduler design (see [6] for details). First, the constraints on available physical resources, include

- The HSDPA reserved power, which is reserved by the RNC to the HSDPA system. The selected MCS, which determines the user data rate, depends directly on the power reserved for HSDPA.
- The number of spreading codes reserved by the RNC to HSDPA since this limits the available MCSs.
- The number of available HS-SCCH channels, which represent an overhead power. In addition, code multiplexing the users in the same TTI to increase the scheduler flexibility depends on the number of available HS-SCCH reserved by the RNC.

Second, there are radio channel conditions, reported on the CQI each TTI, so that the scheduler selects the user with favorable channel conditions. And third, QoS constraints, include

- Amount of data in the node B buffer, which represents a significant parameter, especially for services with limited tolerable jitter.
- Guaranteed bit rate for certain services, such as streaming.
- Tolerated delay jitter.
- Services priority, which allows the node B to prioritize flows relative to other flows and is indicated by the parameter scheduling priority indicator (SPI) set by the RNC.
- Allocation and retention priority (ARP), which allows the node B to determine the priority of a bearer relative to other UMTS bearers.
- Discard timer, which is of significant relevance for streaming services. In fact, the data flows transferred from the RNC to the MAC-hs in the node B, through the Iub interface, are in the form of MAC-d PDUs. On arrival at the node B, a QoS discard timer is set on these

PDUs to limit their maximum queuing. If a given MAC-d PDU is not served before the timer expiration, it is discarded. This parameter should be introduced in the tolerated delay jitter constraint.

■ HARQ status knowledge, where the scheduler should know if the packet to transmit is a new or a retransmitted packet. According to this knowledge, a priority indicator can be introduced to prioritize the retransmitted packets to satisfy the QoS delay constraints.

4.8.2 Selected Scheduling Algorithms

As already mentioned, many scheduling proposals have been analyzed in the literature to share a channel in a timely manner. In this section, the four most popular and relevant schedulers are described: the round robin (RR), the fair throughput (FT), the Max C/I, and the proportional fair.

4.8.2.1 Round Robin

The RR algorithm allocates the channel to the users in a cyclic order offering fair time resource sharing among users. Since this scheduler ignores the radio channel conditions, it does not provide a fair throughput between users. The absence of the scheduling adaptation to the short-term channel variations counteracts the fast link adaptation introduced in HSDPA. Consequently, this scheduler provides low cell throughput. The only advantage of using this algorithm is its simplicity to implement.

4.8.2.2 Fair Throughput

The FT scheduler allocates the HS-DSCH channel to the users to achieve fair sharing of the entire cell throughput among users. During each TTI, the channel is allocated to the user with the lowest average received data rate over previous TTIs. This algorithm provides fair resource sharing but does not exploit the instantaneous channel information. Therefore, this scheduler neglects and defeats the fast link adaptation of HSDPA and results in poor average cell throughput.

4.8.2.3 Max C/I

The Max C/I scheduler is perfectly suited to adapt quickly to the instantaneous channel variations. During each TTI, the HS-DSCH channel is allocated to the user having the best channel conditions. In fact, the node B uses the CQI reported by the link adaptation procedure and allocates the HS-DSCH channel to the user with the best SIR. In the ideal situation when channel conditions of the users present the same statistics, this strategy maximizes the total capacity of the system and the throughput of individual users. In reality, the statistics are not symmetrical since the users can be closer to the base station with a better average SIR or at the cell border with relatively bad conditions, stationary or moving at high speed, or in a rich scattering environment or with no scatterers around them. Therefore, by using the Max C/I strategy in practice, the channel is always allocated with higher-order MCS (i.e., higher average transmission rate) but induces starvation of users with relatively bad channel conditions. Consequently, this algorithm maximizes the cell capacity but presents a problem of fairness between users especially for users at the cell border. In addition, the QoS constraints (e.g., the throughput in a given time scale and not the long-term average throughput) of different services are not considered in this scheduler, which can have a drastic effect on higher layers such as TCP or on certain services such as streaming.

4.8.2.4 Proportional Fair

As explained previously, the FT algorithm performs a fair sharing of the average cell throughput among users independently of the channel conditions. This causes loss in average throughput per user and per cell. The Max C/I strategy prioritizes users with good channel conditions and thus maximizes the cell throughput but at the expense of price of users at the cell border. A problem of fairness is therefore observed. To achieve a tradeoff between fairness and efficiency, the PF strategy has been proposed and analyzed in [22, 23]. It consists of transmitting to the user with the highest data rate relative to its current mean data rate. During each TTI, the channel is allocated to the user having $\max(r/S)$, with r being the transmission rate in the current TTI according to the transmission scheme selected and S the average bit rate transmitted in the previous TTIs and evaluated through an exponentially weighted low-pass filter. S is evaluated through equation (4.1):

$$S(n+1) = \begin{cases} \left(1 - \frac{1}{t_c}\right) S(n) + \frac{1}{t_c} r(n) & \text{if the TTI } n \text{ is allocated to the user} \\ \left(1 - \frac{1}{t_c}\right) S(n) & \text{elsewhere.} \end{cases} \quad (4.1)$$

This strategy keeps track of the average throughput S of each user in a past window of length t_c, where t_c is a parameter varying between 800 and 1000 in general.

Note that the PF can be regarded as a compromise between Max C/I and FT strategies. For users with the same channel conditions (i.e., the same transmission rate r), the HS-DSCH channel is allocated to the user having the lowest average throughput $(\max(1/S))$, which corresponds to the FT strategy. However, for users with the same average received throughput (i.e., the same S), the HS-DSCH channel is assigned to the user having the highest r (i.e., the best SIR), which corresponds to the Max C/I strategy.

In the literature, there are several versions of the PF. Some versions propose to evaluate the mean bit rate R through an exponentially smoothed average when others increase the influence of the instantaneously transmission rate by using the user selection condition [42,43]: $max(r^c/S)$, where c is a parameter depending on the channel conditions. These algorithms increase the fairness at the expense of cell throughput and vice versa. The selected algorithm is entirely up to the implementer and the operator strategy. The PF algorithm, as defined in [22], can be considered a reasonable trade-off between fairness and capacity.

4.9 HSDPA Modeling and Cell Throughput

This section presents a simple and basic semianalytical model allowing for the estimation of the cell throughput. This model captures the effect of HARQ Chase combining, AMC, and scheduling.

The analysis is broken down into several steps, in which the first phase addresses HARQ and the next steps cover modeling of AMC and scheduling. At the end of this three-step analysis, a model for each scheduling algorithm is reported.

4.9.1 HARQ

In [44], it was shown that the average number of transmissions N_s due to HARQ can be evaluated using the following expression:

$$\overline{N_s} = \frac{1 + P_e - P_e P_s}{1 - P_e P_s}, \tag{4.2}$$

where P_e is the BLER and P_s is the probability of errors after soft combining two successive erroneous transmissions using the Chase combining algorithm [17].

4.9.2 AMC

To track the variation of the channel conditions, AMC is used in HSDPA where a MCS (a modulation order M, a coding rate τ, and a number of HS-DSCH codes N) is selected on a dynamic basis according to the value of SIR. Note that for N, the maximum number of available HS-DSCH codes is 15. Let $SIR = \gamma_{Rake}$ for a Rake receiver and k_{mcs} be the probability of selecting a modulation and coding scheme mcs. The probability of selecting a given *mcs* combination can be expressed as $k_{mcs} = Prob(SIR \geq \gamma_{mcs})$ for the highest-order transmission scheme and $k_{mcs} = Prob(\gamma_{mcs} \leq SIR < \gamma_{mcs+1})$ for the other transmission schemes, where γ_{mcs} is the target *SIR* of the modulation and coding scheme *mcs*. The probability $P(\gamma_{mcs} \leq SIR < \gamma_{mcs+1}) = P(SIR \leq \gamma_{mcs+1}) - P(SIR \leq \gamma_{mcs})$ can be evaluated, for each user, through simulation.

Once the probability k_{mcs} has been extracted from the simulations, the cell throughput can be estimated through analytical expressions provided for each scheduler. These expressions are the subject of the next section.

4.9.3 Scheduling

As fast scheduling is one of the key techniques used in HSDPA, the user bit rate and the cell capacity will depend on the used scheduler. To assess achievable performance, general analytical expressions for cell capacity and user bit rate are specialized to four schedulers: FT, RR, Max C/I, and PF.

4.9.3.1 Round Robin Scheduler

In the RR scheduler, the channel is shared equally between users—the same number of TTIs is allocated to each user. If N_u is the number of users in the cell, then the probability that a TTI is allocated to a given user is $1/N_u$. Hence, the mean bit rate of user i is given by [27–29]

$$R_i = \frac{1}{N_u} \sum_m \frac{R_m k_{m,i}}{N_{s,i}}$$

$$= \frac{1}{N_u} \sum_m k_m \frac{W}{SF} \frac{(N \log 2(M)\tau)_{m,i}}{N_{s,i}}, \tag{4.3}$$

where R_m is the bit rate of transmission scheme m during a TTI, $N_{s,i}$ is the average number of packet transmissions of user i due to HARQ, W is the chip rate, and SF is the spreading factor. Note that $k_{m,i}$ varies with the mobile position. The cell throughput in this case is given by

$$tb = E\left(\sum_{i=1}^{N_u} R_i\right). \tag{4.4}$$

4.9.3.2 Fair Throughput Scheduler

The FT scheduler allocates a fixed bit rate to users independently of channel condition and mobile position. In this case, the cell capacity can be estimated as follows [44]

The HS-DSCH aggregate channel flow in symbols/sec is given by

$$R_s = \frac{W}{SF}.$$ (4.5)

Since the HS-DSCH channel is shared by several users in a given time T (i.e., channel observation time) we have

$$\frac{\sum_{i=1}^{N_u} Rs_i T_i}{T} = \frac{W}{SF},$$ (4.6)

where Rs_i is the throughput of each user in symbols/sec and T_i is the connection duration. The modulation and coding scheme changes during the transfer of the packet calls. Hence, the previous equation leads to

$$\frac{\sum_{i=1}^{N_u} R_i T_i \sum_m \frac{k_m}{(N log 2(M)\tau)_{i,m}}}{T} = \frac{W}{SF},$$ (4.7)

where R_i is the service bit rate. Due to the effect of HARQ, $N_{s,i}$ packets are transmitted instead of one packet, having all the same modulation and coding scheme. Hence, the number of users N_u in the cell can be evaluated using the following equation:

$$\frac{\sum_{i=1}^{N_u} R_i N_{s,i} T_i \sum_m \frac{k_m}{(N log 2(M)\tau)_{i,m}}}{T} = \frac{W}{SF}.$$ (4.8)

Equation (4.8) must be solved for N_u to derive the overall aggregate cell throughput as

$$th = E\left(\frac{\sum_{i=1}^{N_u} R_i T_i}{T}\right).$$ (4.9)

4.9.3.3 Max C/I Scheduler

In Max C/I scheduling, the channel is allocated in each TTI to the user having the best SIR. This scheduler maximizes the cell capacity but does not guarantee any QoS to the user. Users at the border of the cell have always poor channel conditions due to attenuation, interference, and absence of fast power control, and they experience low bit rate.

If N_u is the number of users in the cell, the probability that a TTI is allocated to user i is

$$pr(i) = Prob(SIR_i > SIR_j \text{ for } j = 1.. N_u \text{ and } j \neq i).$$ (4.10)

Since the analytical evaluation of this probability is unwieldy, it is basically evaluated through simulation. Given this probability, the bit rate of user i is provided by [27–29]

$$R_i = pr(i) \sum_m \frac{R_m k_{m,i}}{N_{s,i}}$$

$$= pr(i) \sum_m k_m \frac{W}{SF} \frac{(N \log 2(M)\tau)_{m,i}}{N_{s,i}}. \tag{4.11}$$

The cell throughput in this case is

$$th = E\left(\sum_{i=1}^{N_u} R_i\right). \tag{4.12}$$

4.9.3.4   Proportional Fair Scheduler

The PF scheduler is a compromise between Max C/I and FT. In each TTI, the channel is allocated to the user having $\max(r/S)$ where r is the transmission rate in this TTI—according to the transmission scheme selected—and S is the mean bit rate transmitted in previous TTIs.

A detailed model of the PF is difficult to derive since during each TTI the probability to allocate the channel to a user i depends on all the previous TTIs. To circumvent this difficulty, current literature instead models the asymptotic characteristics of the PF scheduler by considering the condition $\max(r/S)$ as equivalent to $\max(SIR/S)$ and S as the infinite limit of the instantaneous received bit rate [25–29]. According to each SIR value, there is a given possible transmission rate r in a given TTI; in fact, r values correspond to a range of SIRs. Since the number of MCS schemes is high (equal to 30 [11]), this hypothesis is still a good approximation, and it can be seen as an asymptotic study of the PF scheduler. Hence, if N_u is the number of users in the cell, the probability that a TTI is allocated to a given user i can be evaluated using

$$pr(i) = Prob\left(\frac{SIR_i}{S_i} > \frac{SIR_j}{S_j} \text{ for } j = 1..N_u \text{ and } j \neq i\right). \tag{4.13}$$

Consequently, the bit rate achieved by user i is obtained through

$$R_i = pr(i) \sum_m k_m \frac{W}{SF} \frac{(N \log 2(M)\tau)_{m,i}}{N_{s,i}}. \tag{4.14}$$

The cell throughput in this case is given by

$$th = E\left(\sum_{i=1}^{N_u} R_i\right). \tag{4.15}$$

4.9.4 Results

The analytical models presented in Sections 4.9.1, 4.9.2, and 4.9.3 allow for an evaluation of the cell throughput and bit rate of each user for each of the four schedulers: RR, FT, Max C/I, and PF. To compare these four schedulers—that is, to determine the best trade-off between fairness and capacity—results obtained from the analytical models should provide the cell throughput and the achieved user bit rate especially for users with unfavorable radio channel conditions at the cell border, which serves as indication of fairness. Figures 4.13 and 4.14 depicts, respectively, the average cell throughput and the cumulative distribution function (cdf) of a user situated 900 m from the node B. Note that the cell radius is 1000 m, and eight simultaneous users are active in the cell. Results show that the Max C/I scheduler achieves higher cell capacity than other schedulers (approximately 3 Mbps for 15 available HS-DSCH channel codes) without any fairness guarantee: the user at the cell border gets a bit rate up to 50 kbps instead of 370 kbps. The FT scheduler presents lower cell throughput than other schedulers (approximately 1.7 Mbps for 15 available HS-DSCH channel codes) with a perfect fairness: approximately 215 kbps per user. The PF scheduler represents a good compromise between fairness

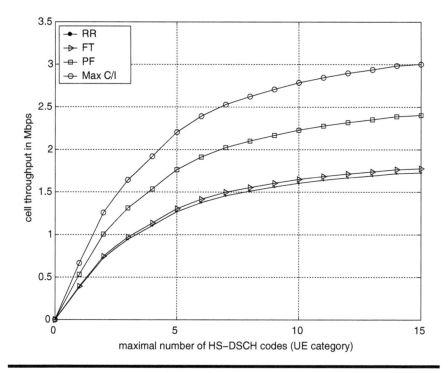

Figure 4.13   HSDPA cell throughput for different scheduling algorithms.

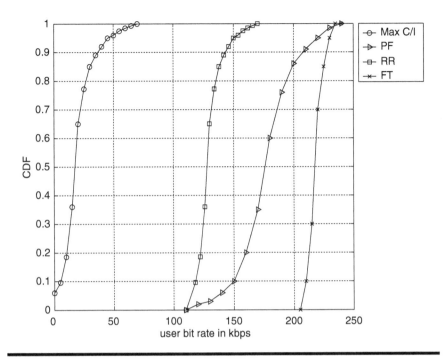

Figure 4.14 Example of the CDF of a user bit rate for different scheduling algorithms.

and capacity. It achieves a cell throughput approximately equal to 2.4 Mbps while guaranteeing an acceptable fairness: The bit rate of the user at the cell border varies between 120 kbps and 220 kbps with an average value equal to 170 kbps).

Note that more detailed descriptions of the models described in Sections 4.9.1, 4.9.2, and 4.9.3 are presented in [27], where the authors have derived complete analytical expressions of the schedulers characteristics and especially of the probabilities $pr(i)$. This reference provides more analysis and results on the comparison between schedulers in HSDPA. Also, in the literature several studies have been conducted to evaluate the performance of schedulers over time-shared channels, in HSDPA and other equivalent systems. Some examples for further reading are [6,22–26,30–43].

References

1. 3GPP TS 25.308 V6.3.0. 2004. HSDPA Overall Description. Stage 2. Release 6, December.
2. 3GPP TR 25.858 V5.0.0. 2002. High-Speed Downlink Packet Access, Physical Layer Aspects. Release 5, March.

3. 3GPP TR 25.877 V5.1.0. 2002. High-Speed Downlink Packet Access (HSDPA) Iub/Iur Protocol Aspects. Release 5, June.

4. 3GPP TR 25.899 V6.1.0. 2004. High-Speed Download Packet Access (HSDPA) Enhancements. Release 6, September.

5. Holma, H., and A. Toskala. 2004. *WCDMA for UMTS: Radio Access for Third Generation Mobile Communications*, 3rd ed., London: John Wiley and Sons.

6. Ameigeiras Gutierrez, P. J. 2003. Packet Scheduling and QoS in HSDPA. PhD diss., Aalborg University, October, 59.

7. 3GPP TS 25.201 V6.2.0. 2005. Physical Layer—General Description. Release 6, June.

8. 3GPP TS 25.211 V6.6.0. 2005. Physical Channels and Mapping of Transport Channels onto Physical Channels (FDD). Release 6, September.

9. 3GPP TS 25.212 V6.6.0. 2005. Multiplexing and Channel Coding (FDD). Release 6, September.

10. 3GPP TS 25.213 V6.4.0. 2005. Spreading and Modulation (FDD). Release 6, September.

11. 3GPP TS 25.214 V6.5.0. 2005. Physical Layer Procedures (FDD). Release 6, March.

12. 3GPP TS 25.331 V6.7.0. 2005. Radio Resource Control (RRC) Protocol Specification. Release 6, September.

13. Tanner, R., and J. Woodard. 2004. *WCDMA Requirements and Practical Design*. London: John Wiley & Sons.

14. 3GPP TS 25.321 V6.5.0. 2005. Medium Access Control (MAC) Protocol Specification. Release 6, June.

15. Ericsson, N. C. 2001. On Scheduling and Adaptive Modulation in Wireless Communications. Technical Licentiate Thesis, UPPSALA University, Sweden, June.

16. 3GPP TSG RAN WG1#28. 2002. Duration of CQI Measurement. Philips, Seattle, WA, August, Tdoc R1-02-1084.

17. Chase, D. 1985. Code Combining—A Maximum-Likelihood Decoding Approach for Combining an Arbitrary Number of Noisy Packets. *IEEE Transactions on Communications* 33, no. 5:385–93 (May).

18. 3GPP TSG RAN WG1#18. 2001. Clarifications on Dual-Channel Stop-and-Wait HARQ. Motorola, Tdoc R1-01-0048.

19. 3GPP TSG RAN WG1#18. 2001. Considerations on HSDPA HARQ Concepts. Nokia, Tdoc R1-01-0007.

20. 3GPP TSG RAN WG1#20. 2001. Further Buffer Complexity and Processing Time Considerations on HARQ. Nokia, Busan, Korea, May, Tdoc R1-01-0553.

21. Bestak, R. 2003. Reliability Mechanisms (Protocols ARQ) and Their Adaptation in 3G Mobile Networks. PhD diss., ENST, Paris, December, http://pastel.paristech.org/archive/00000514/01/RBestakThese.pdf.

22. Holtzman, J. M. 2000. CDMA Forward Link Waterfilling Power Control. In Proc. of the IEEE Vehicular Technology Conference (VTC), 3:1663–7. (May).

23. Jalali, A., R. Padovani, and R. Pankaj. 2000. Data Throughput of CDMA-HDR: A High Efficiency—High Data Wireless Personal Communication System. In Proc. of the 50th Annual IEEE VTC 3: 1854–8, Tokyo, May.

24. Elliott, R. C., and W. A. Krzymien. 2002. Scheduling Algorithms for the CDMA2000 Packet Data Evolution. In Proc. of the 52nd Annual IEEE VTC, 1:304–10 (Fall).

25. Holtzman, J. M. 2001. Asymptotic Analysis of Proportional Fair Algorithm. In Proc. of the 12th IEEE International Symposium on Personal, Indoor and Mobile Radio Communications (PIMRC) 2:F-33-7 (September).

26. Borst, S. 2003. User-Level Performance of Channel-Aware Scheduling Algorithms in Wireless Data Networks. In Proc. of the 22nd IEEE Conference on Computer Communications (INFOCOM), 1:321–31 (March-April).

27. Assaad, M., and D. Zeghlache. 2006. Cross Layer Design in HSDPA System. *IEEE Journal on Selected Areas in Communications*, 24, no. 3:614–25 (March).

28. Assaad, M., and D. Zeghlache. HSDPA Performance under Nakagami Fading Channel. *IEEE Transactions on Wireless Communications*, forthcoming.

29. Assaad, M., and D. Zeghlache. 2005. Scheduling Performance in HSDPA System under Rice Fading Channel. *IEEE Transactions on Vehicular Technology*, forthcoming.

30. Kolding, T., F. Frederiksen, and P. Mogensen. 2002. Performance Aspects of WCDMA Systems with High-Speed Downlink Packet Access (HSDPA). *IEEE Semiannual Vehicular Technology Conference* 1:477–81 (Fall).

31. Liu, Xin, Edwin K. P. Chong, and Ness B. Shroff. 2001. Opportunistic Transmission Scheduling with Resource-Sharing Constraints in Wireless Networks. *IEEE Journal on Selected Areas in Communications*, 19:2053–64 (October).

32. Hosein, P. 2002. QoS Control for WCDMA High Speed Data. Paper presented at the 4th International Workshop on Mobile and Wireless Communications Network, September, 169–73.

33. Fragouli, C., V. Sivaraman, and M. Srivastava. 1998. Controlled Multimedia Wireless Link Sharing via Enhanced Class-Based Queueing with Channel-State Dependent Packet Scheduling. In Proc. of INFOCOM 98, 2:572–80 (March).

34. Lu, S., and V. Bharghavan. 1999. Fair Scheduling in Wireless Packet Networks. *IEEE/ACM Trans. Networking* 7, no. 4:473–89.

35. Ng, T., S. Eugene, I. Stoica, and H. Zhang. 1998. Packet Fair Queueing Algorithms for Wireless Networks with Location-Dependent Errors. In Proc. of INFOCOM 98, 2:1103–11 (March).

36. Ramanathan, P., and P. Agrawal. 1998. Adapting Packet Fair Queueing Algorithms to Wireless Networks. In Proc. of the 4th ACM International Conference on Mobile Computing and Networking (MOBICOM 98), Dallas, 1–9 (October).

37. Gomez, J., A. T. Campbell, and H. Morikawa. 1999. The Havana Framework for Supporting Application and Channel Dependent QoS in Wireless Networks. In Proc. of ICNP 99, 235–44, November.

38. Wang, L., Y.-K. Kwok, W.-C. Lau, and V. K. N. Lau. 2002. *Channel Capacity Fair Queueing in Wireless Networks: Issues and a New Algorithm*. New York: ICC.

39. Andrews, M., K. Kumaran, K. Ramanan, A. Stolyar, P. Whiting, and R. Vijayakumar. 2001. Providing Quality of Service over a Shared Wireless Link. *IEEE Communications Magazine* 39, no. 2:150–4 (February).
40. Xu, L., X. Shen, and J. Mark. 2002. Dynamic Bandwidth Allocation with Fair Scheduling for WCDMA Systems. *IEEE Wireless Communications* 9, no. 2:26–32 (April).
41. Shao, H.-R., C. Shen, D. Gu, J. Zhang, and P. Orlik. 2003. Dynamic Resource Control for High-Speed Downlink Packet Access Wireless Channel. Paper presented at the 23rd IEEE International Conference on Distributed Computing Systems Workshops, May, 838–43.
42. Kim, H., K. Kim, Y. Han, and J. Lee. 2002. An Efficient Algorithm for QoS in Wireless Packet Data Transmission. In Proc. of the IEEE International Symposium on PIMRC, 5:2244–8 (September).
43. Aniba, G., and S. Aissa. 2004. Adaptive Proportional Fairness for Packet Scheduling in HSDPA. *IEEE Global Telecommunications Conference* 6:4033–7 (November–December).
44. Assaad, M., and D. Zeghlache. 2003. On the Capacity of HSDPA System. *IEEE Globecom Conference* 1:60–64, San Francisco, CA, December 1–5.

Chapter 5

Applications and Transport Control Protocol

The emerging growth in application services over wired and wireless systems in recent years results from the phenomenal evolution in Internet and wireless systems. This growth has fostered the development of advanced Internet applications and services and has generated a number of additional challenges in controlling congestion in networks to provide adequate QoS to applications.

Traditionally, applications can be classified into real-time and nonreal-time services according to their transmission delay requirements. The RT services impose strict delay requirements on end-to-end connections. These severe delay constraints prevent the use of link level retransmissions (ARQ) and of error recovery mechanisms at higher layers. Therefore, the use of reliable transport protocol such as TCP, commonly used to control congestion in the Internet, is dismissed in this case. RT services are usually conveyed by unreliable transport protocols such as the user datagram protocol (UDP) and are consequently subject to transmission errors.

NRT services, on the other hand, are characterized by high sensitivity to errors, though with much lower delay constraints than RT services. Link retransmissions and end-to-end recovery mechanisms are suitable for managing NRT traffic so that data is delivered to upper layers without errors. These services can be carried by reliable transport protocols such as transport control protocol.

This book considers only NRT services conveyed by reliable transport protocols, especially TCP. This chapter presents the TCP protocol with an emphasis on protocol presentation, flow control, and modeling. The chapter also serves as background for Chapter 6, which addresses interactions between radio link retransmission mechanisms and TCP. When TCP is used over wireless it commonly mistakes retransmissions over the air link for congestion in the fixed network segments.

5.1 UDP Services

The user datagram protocol [1–3] is a transport protocol provided by the IP protocol stack and is used in data networks to carry various application services. UDP transmits segmented data, provided by the application layer, as independent datagrams. UDP is a datagram-oriented protocol. In addition, UDP is connectionless and does not implement connection establishment and connection termination [4]. There is no flow control to generate buffer overflow in UDP. UDP is not a reliable transport protocol, as it does not implement any retransmission or recovery mechanisms. Hence, errors or datagram losses during transmission are not corrected.

UDP is suitable for applications that transmit short messages through the network, such as Simple Network Management Protocols (SNMP) [5] and routing applications, like Routing Information Protocol (RIP) [6–8]. Another type of application that uses UDP is the multicasting service (e.g., webcasting) when the same information is sent to several destinations. UDP is suitable for these kinds of services since it is a connectionless protocol. UDP does not suffer from scalability problems encountered with TCP.

The major use of UDP is in the transport of real-time multimedia services, due essentially to the tight delay constraints that characterize this type of application. Since no error-recovery mechanisms are used in UDP, robustness has to be built into the application and the source encoder. In this context the Real-Time Protocol (RTP) [9] was developed for real-time interactive communications (e.g., streaming). RTP includes a sequence number field to the RTP packets to detect packet losses and to restore the packet sequence. In addition, RTP frames the data to be delivered to UDP, allows the synchronization at the receiver, and removes the delay jitter caused by the network.

5.2 TCP Services

On the contrary, TCP [10] is a reliable transport protocol widely developed and implemented in current data networks. TCP performs end-to-end recovery to ensure the transmitted data reaches its destination without errors. In addition, the transmission rate and the bandwidth occupied by each connection are controlled by a signal feedback from the receiver and by a sliding transmission window scheme.

TCP offers several services to the application layer, including [2–4] reliability, end-to-end semantic, connection oriented, full duplex, and streaming.

TCP is a reliable protocol. It ensures that each transmitted packet is received correctly by the destination. The received packets are stored in a buffer and are delivered in sequence to the upper layers. No duplication occurs during this procedure. The erroneous packets are retransmitted using an end-to-end recovery mechanism similar to ARQ. This protocol is adapted to applications that have high sensitivity to errors and that tolerate delays such as file transfer and access to the WWW. The applications that rely on TCP do not have to worry about packet errors or lost data. TCP is in charge of error-free delivery of ordered data to the upper layers, unlike UDP where the applications have to be robust enough to deal with losses in the received data.

Note that the error-recovery mechanism is provided by the end-to-end semantic of the TCP protocol. Even though the TCP packets pass through various network nodes, only the destination acknowledges the error-free reception of the packets. In case of errors or lost data, a negative acknowledgment is sent to the transmitter so that the lost data can be retransmitted. The end-to-end semantic ensures the reliability of the TCP.

TCP is a connection-oriented protocol. In fact, each application relies on a specific TCP connection, which is identified by a source TCP port number, destination TCP port number, source IP address, and destination IP address. Each connection starts with establishing a TCP connection and ends with terminating a connection, as explained later in this chapter.

The connection established by TCP between two application processes—sending and receiving processes—is a full duplex connection. This means that both of the application processes can send data over the same TCP connection at the same time.

Finally, the TCP packets transmitted over a TCP connection are dependent. Unlike UDP, TCP stores the correctly received packets in a buffer, extracts the applications from each packet, and delivers them to the application as a stream. Therefore, TCP is stream oriented. Note that the stream-oriented abstraction does not mean that TCP is suitable for streaming applications. Rather, the RTP–UDP combination is often used for this kind of applications.

These services provided to the applications make TCP the most popular transport protocol used in the Internet. Applications such as e-mail, remote login (TELNET), WWW, and file transfer (e.g., images, video clip, computer programs) rely on TCP due to their requirements. Let us briefly describe these applications.

5.2.1 World Wide Web

The most popular application used over the Internet is the World Wide Web, which relies on the Hypertext Transfer Protocol (HTTP) [11,12] to download objects from a Web site. Each Web page is written using the Hypertext Markup Language (HTML). HTML divides each page into primary and secondary files. The secondary files contain the Web-page images, whereas the primary file contains the textual contents of the page and references to secondary files as well as to other Web pages.

Two versions of the HTTP protocol are used: HTTP 1.0 and HTTP 1.1. The main difference is that HTTP 1.1 considers the persistent connection as default in any HTTP connection [12]. In fact, with HTTP 1.0 each primary or secondary file requires a TCP connection that generates an overhead of open and closed connections. In the case of Web browsers allowing one connection at a time, this can generate inefficiency due to the time lost during the multiple TCP establishments and terminations. Current Web browsers solve this problem by allowing several connections at a time at the risk of increasing congestion. However, with HTTP 1.1 this problem can be avoided. The persistent connection feature of this version allows the retrieval of all files with a single TCP connection. In addition, with HTTP 1.1 the user can send multiple requests without waiting for a response for each request before sending another. This procedure, called requests pipelining, resembles the transmission with a sliding window (explained in the ARQ section) that increases the protocol efficiency and reduces the data transfer delay between two HTTP application processes.

The basic packets exchange during an HTTP 1.1 application, which relies on a TCP connection, is presented in Figure 5.1 (for more details see [13]). To download a Web page from a server, the application layer, in host A, sends an HTTP request containing the universal resource locator (URL) of the page to the TCP layer. Before sending the request to the concerned server, the TCP establishes a connection with this server. This connection establishment consists of synchronization message (SYN) from host A to host B (server) and an acknowledgment (ACK) to this message from host B followed by an acknowledgment from host A. Once host B receives the HTTP request, it reads the HTML primary file and transfers it as a stream to the TCP layer. TCP segments the received stream into packets whose transmission is ruled by the TCP protocol (e.g., slow start, congestion avoidance, sliding window). The TCP protocol is explained later in this chapter. Once host A's TCP layer receives the entire primary file, it delivers the data in sequence to the upper layer. Host A's application layer parses the file and extracts the URLs of the secondary objects. The application layer sends several HTTP requests, containing the URL of the secondary objects, and requests are sent before receiving a reply to previous requests. Note that all secondary files are transferred using the same TCP connection.

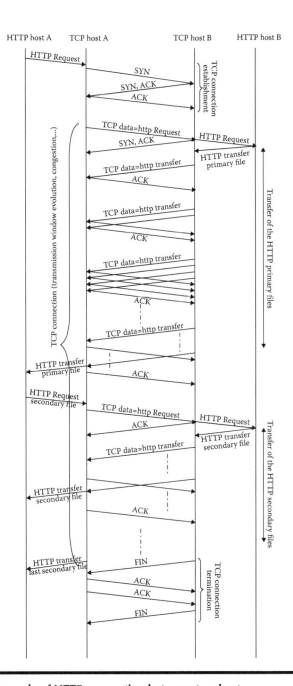

Figure 5.1   Example of HTTP connection between two hosts.

Once host A parses the entire secondary file and reconstructs the page, the reading time starts. The reading time depends on the human behavior and the page contents. Even though the reading time starts, the TCP connection is still maintained, since a persistent connection is used, until host A or host B closes the connection. Therefore, an HTTP session—several Web sessions and reading times—relies on the same TCP connection, but this requires huge resources and generates inefficiency, especially in the case of long reading times (for more information about parsing time see [3,14]).

To deal with the wasted resources generated by the persistent connection, the Apache HTTP server (version 1.3) defines a holding time limited to 15 seconds [15]. Once the connection is established, the server starts a timer, and the connection is terminated if the server or the user closes it or if the timer expires. According to [16], 50 percent of the reading times have a duration more than 15 seconds, which implies that in the majority of HTTP sessions (version 1.3) several TCP connections are required.

5.3 TCP

As mentioned already, TCP is a connection-oriented reliable transport protocol. To provide useful services (described in Section 5.2), this protocol performs four major functions [2–4]: data segmentation, flow control, error control, and congestion control. In this section, the TCP protocol mechanisms utilized during data transmissions are described.

5.3.1 Connection Establishment and Termination

The connection-oriented service of TCP allows the data transmission of each application over a specific connection. Before starting the transmission, TCP proceeds to a connection establishment called three-way handshaking (see Figure 5.2). This connection establishment consists of a SYN segment with an initial sequence number sent by the client to the server lodging the file or the page to be transferred. The server answers to this SYN segment by a TCP segment containing the SYN, a sequence number, an acknowledgment to SYN, and an ACK number. The ACK number indicates the sequence number of the previously received data. It tells the client the sequence number of the next segment that should be transmitted. The sequence number in the server answer indicates the start sequence number in the server; the client and the server may use different sequence numbers. The three-way handshaking is terminated by an acknowledgment message from the client containing an ACK number equal to the server sequence number.

Once the data transmission over TCP is finished, or the holding time is expired as explained in the previous section, the TCP protocol closes the

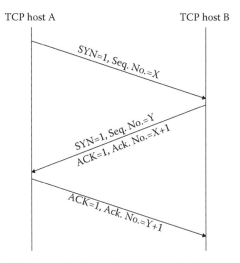

Figure 5.2   Example of TCP connection establishment (three-way handshake).

connection. This connection termination is called four-way handshaking (see Figure 5.3). To better describe this procedure, consider when the client decides the close the connection. In this case, the client sends a FIN segment asking to close the connection. This segment is acknowledged by the server and the connection in the client-server direction is closed. Once the

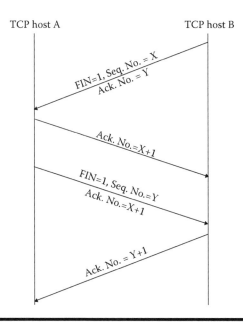

Figure 5.3   Example of TCP connection termination.

server is ready to close the communication, it sends a FIN message to the client, which answers by an acknowledgment segment ACK. This closes the connection in the server–client direction.

5.3.2 TCP Segmentation

The data delivered from the application layer is segmented at the TCP level into chunks or segments. Each TCP segment consists of a 20-byte header, and up to 40 bytes if TCP options are used, which is followed by a variable data payload size. The TCP segment format is depicted in Figure 5.4.

According to [17], the header of a TCP segment contains the following fields:

- Source port number (16 bits), which is used to identify a TCP application at the source host.
- Destination port number (16 bits), which is used to identify a TCP application at the destination host. Note that the source and the destination port numbers are useful to multiplex different TCP connections over the same TCP protocol process [17].
- Sequence number (32 bits) of the first data octet in this segment, except when SYN is present. If SYN is present the sequence number is the initial sequence number, and the first data octet is this sequence number plus 1.
- Acknowledgment number (32 bits), which is used to acknowledge the correctly received data and contains the value of the next sequence number the sender of the segment is expecting to receive.

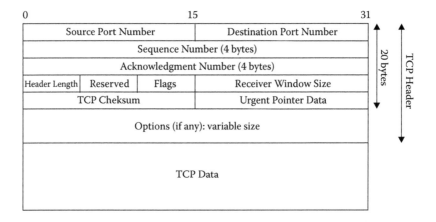

Figure 5.4 Structure of TCP header.

- Header length (4 bits), which indicates the number of 32-bit words in the TCP header showing where the data begins. Since the options field in the TCP header has a variable size, the header size is variable, and knowing the header length is of interest to specify the start of the data-block size.
- Reserved unused (6 bits), which is reserved for future use.
- Flags (6 bits), which are used to relay control information among TCP peers and include the SYN flag, used when establishing a TCP connection; the FIN flag, used in the termination phase of a TCP connection; the ACK flag, set any time the acknowledgement field is valid, implying that the receiver should pay attention to it; the URG flag, signifying that this segment contains urgent data—when this flag is set, the urgent pointer field indicates where the nonurgent data contained in this segment begins; the PUSH flag, signifying that the sender invoked the push operation, which indicates to the receiving side of TCP that it should notify the receiving process of this fact; and the RESET flag, signifying that the receiver has become confused and so wants to abort the connection.
- Receiver window size (16 bits), which is "the number of data octets beginning with the one indicated in the acknowledgment field the sender of this segment is willing to accept" [17]. This field represents the available buffer space at the receiver and allows the sender to better adapt the TCP flow control.
- Checksum (16 bits), which covers the TCP header, the TCP data, the IP source and destination addresses, and the IP header length. This field is mandatory and must be calculated by the sender and then verified by the receiver to protect the TCP segment against misrouted segments.
- Urgent pointer data (16 bits). This pointer is used when urgent data is transmitted and when TCP allows other data transmissions to escape. In this case, the urgent pointer field is needed to point to the sequence number of the octet following the urgent data.
- Options (variable), which can be added to the TCP segment to improve the protocol operation. An example of these options is the maximum segment size option (MSS). Each end of the connection normally specifies this option on the first segment exchanged. It specifies the maximum-sized segment the sender wants to receive. Another option that can be used in this field is the timestamp transmitted over 8 bytes. When used this option allows a more accurate estimation of the round-trip time (RTT) to improve the TCP efficiency.

Concerning the application data field, the size of the largest data payload is MSS, which is negotiated between the peer TCP entities during the

Table 5.1 Common MTUs in the Internet from [18]

Network	Maximum Transmission Unit (MTU) in Bytes
Official Maximum MTU	65535
16Mb IBM Token Ring	17914
IEEE 802.4	8166
IEEE 802.5 (4 Mb max)	4464
FDDI	4352
Wideband Network	2048
IEEE 802.5 (4 Mb recommended)	2002
Ethernet	1500
Point-to-Point (Default)	1500
IEEE 802.3	1492
SLIP	1006
ARPANET	1006
X.25	576
DEC IP Portal	544
NETBIOS	512
ARCNET	508
Point-to-Point (Low Delay)	296

three-way handshaking of the TCP connection. MSS is selected in such way to avoid any segmentation or reassembling of IP packets within the connection path. The upper size of MSS is delimited by the size of the maximum transfer unit (MTU) at the IP level. Note that each MTU contains an MSS block and a TCP/IP and is the largest packet a network can transmit without fragmentation. The values of MTU in typical data network are presented in Table 5.1 [18]. Note that the typical values of MSS are 256, 512, 536, or 1460 bytes. In case one of the peer TCP entities does not receive a MSS option from another peer entity, a default value of 536 bytes is selected for MSS [3].

5.3.3   Flow Control and Sliding Window Mechanisms

To avoid buffer overflow at the receiver, especially in the case of high-speed transmission, TCP uses flow control.

To increase protocol efficiency, TCP transmits the data segments through the network using a sliding window scheme. Instead of waiting for the acknowledgment of each segment before transmitting another, TCP transmits several segments at a time. The flow control is in charge of determining the window size or number of TCP segments, to be transmitted at a given instant to optimize the system efficiency and the connection throughput without generating congestion in the network or buffer overflow at the receiver. The control of window size is accomplished by selecting at any time the minimum between the congestion window (cwnd) and the advertised window (awnd). The advertised window is the window advertised by the receivers in every segment transmitted to the sender (e.g., server). This advertised window, imposed by the receiver, is upper bounded by the buffer size of the receiver so that no buffer overflow can emerge. The congestion window is in charge of adapting the amount of TCP data transmissions to the network capacity limit. This congestion window is controlled by the slow start and the congestion-avoidance mechanisms as explained in Section 5.3.5. Consequently, the TCP flow control is accomplished by a dynamic transmission sliding window.

5.3.4 Acknowledgment and Error Detection

As mentioned already, TCP is a reliable protocol that assures the error-free reception of transmitted data segments. Therefore, it relies on an acknowledgment from the receiver to confirm the correct delivery of data. This acknowledgment can be sent in an ACK segment (i.e., no data is sent in this segment) or in a data segment in which the process is called piggybacking. In case of errors in the receiving segments, TCP proceeds to a retransmission of the erroneous packets using an ARQ protocol combined with a dynamic transmission window to control the congestion in the connection path.

Different acknowledgment features can be used in TCP; the most important are delayed, cumulative, and duplicate acknowledgment.

Delayed acknowledgment occurs when the receiver does not acknowledge each segment as soon as it is received. The acknowledgment is delayed for a while so that the receiver can acknowledge two or more segments and can reduce the acknowledgment traffic. Cumulative acknowledgment occurs when each acknowledgment, having a given ACK number, is an acknowledgment to all segments having a sequence number up to this ACK number. Duplicate acknowledgment happens when the sender receives the same ACK several times. In fact, when a segment is lost in the network and the following segments, having sequence numbers beyond the expected number, are received correctly by the receiver, TCP buffers these error-free segments and sends an acknowledgment of the received packets so far in sequence. The regeneration of an acknowledgment with the same ACK number—since the segment having a sequence number equal to this ACK number is lost—causes a duplicate acknowledgment. This phenomenon has a direct consequence on the congestion window control explained in Section 5.3.5.

TCP detects packet losses in two ways: triple duplicate or timeout [2,3]. The retransmission mechanisms are different in each case. In the case of triple duplicate, fast recovery algorithms are incorporated in the new versions of TCP (Reno [19] and beyond). An earlier TCP version (Old Tahoe [10]) deals with the packet loss problem through the retransmission timeout only.

The triple duplicate means that the transmitter considers a given segment to be lost if it receives the same acknowledgment message four times. The timeout occurs when a timer expires before an acknowledgment is received. In fact, each time a new segment—or a new sequence of segments—is sent, a retransmission timer starts. The duration of the retransmission timer is called retransmission timeout (RTO) [20]. The value of the RTO indicates the maximum delay for transmission of the segment to the receiver and transmission of the corresponding ACK from the receiver to the transmitter. When the timer expires, the sender considers the earliest unacknowledged packet as lost and retransmits it.

In general, TCP should accommodate varying delays, especially when it is applied in a dynamic environment that experiences a variable RTT. Note that RTT is the delay between the transmission of a segment and the reception of its acknowledgment. From a performance point of view, an optimum RTO value is mandatory to cope with variable RTTs. In fact, a RTO value longer than necessary may increase the application delay if losses are frequent. On the other hand, a RTO value smaller than necessary may generate premature timeouts, resulting in a loss of resource efficiency.

The management of the retransmission timer is described in [20]. The sender implements a RTT estimator to keep track of the RTT variations, the mean value and standard deviation. The mean RTT and the standard deviation estimates are smoothed by a low pass filter, which prevents the possibility of tracking sudden RTT changes. The smoothed RTT estimation and the RTT variation are updated as follows [20]:

$$smoothedRTT(n) = (1 - \alpha) \times smoothedRTT(n - 1) + \alpha \times RTT(n) \quad (5.1)$$

$$smootheddeviation(n) = (1 - \beta) \times smootheddeviation(n - 1) + \beta$$

$$\times |RTT(n) - smoothedRTT(n)|, \quad (5.2)$$

where the values of α and β are 1/8 and 1/4, respectively [20]. Finally, the RTO is set as follows:

$$RTO = smoothedRTT + max(G, 4 * smootheddeviation), \quad (5.3)$$

where G is the server clock granularity. According to Karn's algorithm, RTTs of retransmitted segments are not included in the RTT estimation.

5.3.5 *Congestion Control and Retransmission Mechanism*

The flow control implemented in TCP prevents an overflow at the receiver by adapting the advertised window dynamically to the receiver buffer space. However, this flow control does not cope with the buffer overflows in the intermediate network nodes. To deal with network congestion, congestion control mechanisms have been implemented in TCP. These mechanisms differ from one TCP version to another.

5.3.5.1 *Slow Start*

The principle behind the slow start is to not overload the network with TCP segments transmitted by a sender. The sender starts the transmission with an initial congestion window of one MSS in recent enhancements, the congestion window can be initialized to two or four MSS. Once the sender receives the acknowledgement of the transmitted packet, it increases the congestion window by one MSS. Henceforth, it can transmit two segments at a time instead of one. As the sender receives acknowledgments of the transmitted segments, it increases the congestion window by one MSS for each acknowledgement received (see Figure 5.5). This procedure continues until a loss is detected, either by triple duplicate or a retransmission timeout, or until the window size reaches a threshold called slow-start threshold. Once the slow-start threshold is reached, the slow-start soft state is replaced with the congestion-avoidance soft state, where the growth of the congestion window slows down.

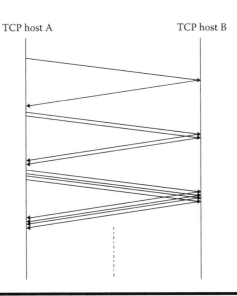

Figure 5.5   Example of TCP slow-start.

The slow-start mechanism can be regarded as a soft phase where the sender probes the buffer space in the network [4]. The successive acknowledgements are interpreted by the sender as an available space in the network to more transmitted packets. Therefore, the sender keeps increasing the window size. On the other hand, the aggressive exponential growth of the congestion window aims at rapidly filling the bottleneck to have a small effect on the throughput performance and to optimize the system efficiency.

5.3.5.2 Congestion Avoidance

As soon as the congestion window reaches the slow-start threshold, the congestion-avoidance soft state takes over the evolution control of the congestion window. During this phase, TCP slows down the probing of the network for more available bandwidth [4]. The congestion-window increase becomes linear: It is increased by one every cycle or RTT. In other words, the congestion window (*cwnd*) gains incrementally by $1/cwnd$ each time an acknowledgment is received. TCP gets out of this congestion phase once a segment loss is detected.

Once a segment loss occurs, the behavior of TCP depends on how this loss is detected by a triple duplicate or a timeout.

The evolution of the congestion window during the slow-start and congestion-avoidance phases is depicted in Figure 5.6.

5.3.5.3 Retransmission Timeout

When a timeout occurs—when the timer expires before receiving an acknowledgment—TCP interprets this phenomenon as a severe congestion in the network. The network is overloaded, and the transmitted segments are lost, which implies retransmission of the segments and a brutal reduction of the congestion window.

The earliest unacknowledged segment is then retransmitted via fast retransmission, and the *RTO* is updated using the update called *back off the timer* [20]:

$$RTO_{new} = 2 \times RTO_{old}. \tag{5.4}$$

In addition, the congestion window *cwnd* is set to the value of the so-called loss window (LW), which in general equals to its initial value (LW = MSS). Subsequently, the slow-start algorithm takes over the evolution control of *cwnd*. The slow-start threshold is also updated as follows [21]:

$$ssthresh = max(FlightSize/2; 2 * \text{MSS}), \tag{5.5}$$

where *FlightSize* represents the amount of TCP segments transmitted but not yet acknowledged.

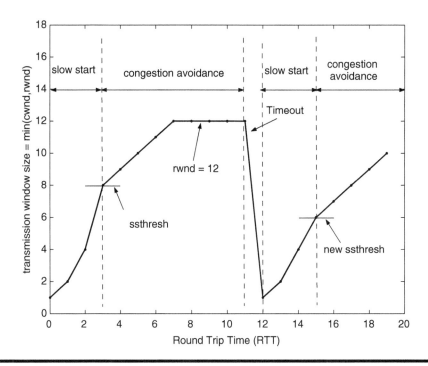

Figure 5.6   Example of TCP congestion window size evolution.

5.3.5.4   Triple Duplicate

When a loss is detected by triple duplicate ACK, TCP interprets this phenomenon as a congestion in the network, and the lost segment is retransmitted via fast retransmission algorithm. Then the control of the *cwnd* evolution depends on the implemented version of TCP.

In Tahoe, TCP reacts as if a timeout occurs. The slow start takes over the *cwnd* control, and RTO and the slow-start threshold are updated according to expressions (5.4) and (5.5).

Since in the case of triple duplicate the sender continues to receive acknowledgments—which is not the case during a timeout—the congestion in the network can be considered less severe than in a timeout. Therefore, to improve the connection efficiency and to deal with data losses, a fast-recovery algorithm has been developed and implemented in the TCP Reno version and in more recent versions.

The fast-recovery state, as implemented in the Reno version, consists of retransmitting the lost segment, restarting the timer with the same RTO [20,21], updating the slow-start threshold according to (5.5), and setting *cwnd* according to the following expression [21]:

$$cwnd = ssthresh + 3 * MSS. \tag{5.6}$$

This update artificially inflates *cwnd* by the number of duplicate acknowledgments—that is, the number of segments that have been received and stored in the receiver buffer. Every subsequently received duplicate acknowledgment indicates that a segment has been received and stored in the receiver buffer. This fast-recovery phase is terminated as soon as a fresh acknowledgment is received. The fresh acknowledgment acknowledges all the segments up to and including the data that was outstanding when the fast recovery started. Once the recovery phase has ended, a new segment can be transmitted by the TCP sender. The congestion window *cwnd* is then set to slow-start threshold, and the congestion-avoidance state takes control of the *cwnd* evolution.

5.4   TCP Modeling

The TCP protocol has been widely studied and discussed in the literature. With the growth of the applications services relying on the TCP/IP protocol stack, the performance of TCP in the currently developed data networks has become a more and more relevant research topic. A large number of analytical TCP models have been developed with the aim of describing the TCP interaction between the TCP protocol and the protocols and the control mechanisms used in a real data network. Even though these models consider a wide variety of TCPs, from TCP Tahoe [22,23] and Reno [23–29] to the new versions such as TCP SACK and Vegas [28–32]. TCP Reno is the version most addressed in the literature, since it is widely implemented in current systems. In addition, developments and modeling of the TCP-friendly congestion control are described in [33–42].

The TCP models proposed in the literature can be divided into four groups [4]: independent packet loss model, random loss model, network model, and control system model. This section presents an overview of these groups of models (further details on TCP modeling can be found, for example, in [4]).

5.4.1   Independent Packet Loss Models

Major approaches have been developed to capture the characteristics of TCP and to evaluate connection throughput and latency time. To develop a mathematically tractable and solvable model, many assumptions are made to simplify the analysis. In most of these models, the TCP performance (i.e., latency and throughput) is described based on the network parameters such as TCP, RTT and packet loss rate. Some of these approaches consider only the triple duplicate without capturing the timeout and its impact on TCP throughput [43,44]. In [43], a periodic loss rate p is assumed, and the steady state throughput is deduced using the following expression:

$$tb = \frac{MSS}{RTT}\frac{K}{\sqrt{p}}.$$

(5.7)

This model is very basic and does not capture the characteristics of TCP in a real network.

In [26-27,44], it was observed that 90 percent of the congestions in TCP result in timeouts. Therefore, for accurate TCP performance predictions the TCP models should include the slow-start phase at the end of each timeout. In addition, the loss model should be random or bursty.

The most popular model that includes the timeout impact was proposed by Padhye et al. in [26,27] and assumes TCP Reno and an independent packet loss model. This model provides a simple equation that can be used to determine the TCP transmission rate according to the packet loss rate, the RTT, and the base timeout value. The analytical approach in [26,27] provides a good prediction of the TCP characteristics in the steady-state phase without considering the connection establishment (i.e., three-way handshake) and the slow-start phase at the beginning of the connection. This model is more appropriate for long-lived TCP connections (i.e., bulk data transfer) where the slow-start phase does not affect the average throughput very much.

According to [45–50], the majority of TCP traffic in the Internet consists of short-lived flows; that is, the transmission comes to an end during the slow-start phase before switching to the congestion-avoidance phase. To predict the TCP performance of short-lived data flows, several models have been proposed [51–53]. These models assume independent packet losses and do not consider the steady-state phase.

To achieve better prediction of the TCP characteristics for short-lived and long-lived data flows, analytical approaches were developed in [54].

This section presents a brief description of the model proposed in [54], which is an extension of the Padhye model [26,27] allowing estimation of the TCP performance for both long- and short-lived data transfer. These two models are the most referenced in the literature.

To evaluate the throughput and the latency time of TCP connections, the TCP behavior is modeled in terms of rounds, where a round is the duration between the transmission of a window of packets and the reception of the acknowledgment of at least one of these packets. In addition, the losses in a round are assumed to be independent of the losses in the other rounds. This model divides the connection duration into five aspects: the three-way handshake time, $T_{handshake}$; the slow-start phase, T_{slow}; the recovery time, $T_{recovery}$, for an eventual loss occurred during the slow start;

the steady-state phase, T_{steady}, for remaining data studied in [26,27]; and the delay acknowledgment time, T_{ack}. The average throughput is given by

$$th = \frac{\text{Packet size}}{E(T_{handshake}) + E(T_{slow}) + E(T_{recovery}) + E(T_{steady}) + E(T_{ack})}. \quad (5.8)$$

The mean value of the connection duration is derived in [54] as follows:

$$T_{handshake} = RTT + T_s \left(\frac{1 - p_r}{1 - 2p_r} + \frac{1 - p_f}{1 - 2p_f} - 2 \right), \quad (5.9)$$

where T_s is the SYN timeout, p_f is the forward packet loss rate along the path from passive opener, to active opener, and p_r is the reverse packet loss rate. If b is the number of TCP segments acknowledged by one acknowledgment, W_1 is the initial window size, W_{max} is the maximum window size, and d is the size of the file in segments, then the average number of segments transmitted during the slow-start phase and the average slow-start latency time are given by

$$E(d_{slow}) = \left(\sum_{k=0}^{d-1} (1 - p)^k p.k \right) + (1 - p)^d . d$$

$$= \frac{(1 - (1 - p)^d)(1 - p)}{p} \quad (5.10)$$

$$E(T_{slow}) = \begin{cases} RTT \left[\log_{\gamma} \left(\frac{W_{max}}{W_1} + 1 + \frac{1}{W_{max}} (E(d_{slow}) - \frac{\gamma W_{max} - W_1}{\gamma - 1} \right) \right] \\ \qquad\qquad\qquad\qquad\qquad \text{When } E(W_{slow}) > W_{max} \\ \\ RTT . \log_{\gamma} \left(\frac{E(d_{slow})(\gamma - 1)}{W_1} + 1 \right) \quad \text{When } E(W_{slow}) \le W_{max}, \end{cases}$$

$$(5.11)$$

where γ is the rate of exponential growth of the window size during slow start and p is the loss rate. $E(W_{slow})$ is given by

$$E(W_{slow}) = \frac{E(d_{slow})(\gamma - 1)}{\gamma} + \frac{W_1}{\gamma}. \quad (5.12)$$

The average recovery time of the first loss can be written using the following expression [54]:

$$E(T_{recovery}) = loss(Q(p, w_{slow})E(z^{TO}) + (1 - Q(p, w_{slow}))RTT), \quad (5.13)$$

where $loss = 1-(1-p)^d$ and $Q(p, w_{slow})$ and $E(z^{To})$ are derived in [26,27] as follows (To is average timeout duration):

$$Q(p, w) = min\left(1, \frac{(1+(1-p)^3(1-(1-p)^{w-3}))}{(1-(1-p)^w)/(1-(1-p)^3)}\right) \qquad (5.14)$$

$$E(z^{To}) = \frac{1+p+2p^2+4p^3+8p^4+16p^5+32p^6}{1-p}To. \qquad (5.15)$$

The remaining data, having a size of $E(d_{ca}) = d-E(d_{ss})$, has an average latency time derived in [26,27] as follows:

$$E(T_{ca}) = \frac{E(d_{ca})}{R(p, RTT, To, W_{max})}, \qquad (5.16)$$

where

$$R(p, RTT, To, W_{max}) = \begin{cases} \dfrac{\frac{1-p}{p} + \frac{W(p)}{2} + Q(p, W(p))}{RTT\left(\frac{b}{2}W(p) + 1\right) + \frac{Q(p, W(p))G(p)To}{1-p}} \\ \qquad\qquad \text{When } W(p) < W_{max} \\[4pt] \dfrac{\frac{1-p}{p} + \frac{W_{max}}{2} + Q(p, W_{max})}{RTT\left(\frac{b}{8}W_{max} + 2 + \frac{1-p}{pW_{max}}\right) + \frac{Q(p, W_{max})G(p)To}{1-p}} \\ \qquad\qquad \text{When } W(p) \geq W_{max} \end{cases} \qquad (5.17)$$

$$W(p) = \frac{2+b}{3b} + \sqrt{\frac{8(1-p)}{3bp} + \left(\frac{2+b}{3b}\right)^2}. \qquad (5.18)$$

5.4.2 Random Loss Model

In the previous section, the packet losses are supposed to be independent. However, the losses in the Internet are often bursty and correlated. Therefore, new models are needed to capture the impact of random correlated losses on TCP performance. Unfortunately, including correlation for losses in the previous model is very complicated. To the best of our knowledge such analytical modeling and analysis do not exist in the literature.

In [55], a simplification of the problem is proposed. Between two losses, the transmission rate T of the TCP connection increases linearly.

The losses occur randomly, and the duration between two losses is a variable S. Since losses are correlated, it can be assumed that S has a general correlation $R_k = E(S_n S_{n+k})$. In addition, the loss process is supposed to be stationary and ergodic. This model does not account for all the TCP characteristics such as the slow start generated by a timeout, the delay acknowledgment, and short-lived data flows. The average throughput can be evaluated using the following expression:

$$E(T) = \frac{1}{RTT\sqrt{p}}\sqrt{0.5\frac{E(S^2)}{E^2(S)} + \sum_{k=1}^{\infty} a^k \frac{R(k)}{E^2(S)}}. \tag{5.19}$$

When the transmission rate is limited frequently by the maximum packet rate the receiver can process (noted P_{rec}), it was shown in [55] that the final average TCP throughput can be approximated by

$$E(T) \simeq P_{rec} - \frac{P_{rec}RTT^2}{8E(S)}. \tag{5.20}$$

5.4.3 Network Model

This modeling approach is developed in [56]. The idea is to predict the performance of all the TCP connections in the network to optimize the network utility—in other words, to find the optimal sharing of the network bandwidth among various connections. It can be regarded as a fair or proportional fair distribution of the network bandwidth among TCP connections. Analytically, this model uses a series of differential equations and Lyapunov functions to find the optimal operation point of the network, which is a vector containing the rate of each TCP connection having a particular route r. If tr_r is the transmission rate of the route r and $\boldsymbol{t^*}$ is the optimal rate vector, it was shown in [56] that proportional fairness is achieved when

$$\sum_{r \in R} \frac{t_r - \boldsymbol{t_r^*}}{\boldsymbol{t_r^*}} \leq 0, \tag{5.21}$$

where R is the group of routes that can be used by the TCP connections. Note that more detailed description of this modeling approach can be found in [4,56].

5.4.4 Control System Model

Analytical models based on the control system are proposed to capture the effect of the random early detection (RED) algorithm on TCP performance. In fact, the losses through the Internet are generated by packets dropping—by one or more nodes—due to the limited buffer space. If an active queueing management (AQM) approach is applied, the node can anticipate the network congestion and can drop some packets. The RED is the most popular packets-discarding algorithm that drops packets according to an average queue length estimation.

According to [57], the average TCP window size $E(W)$ varies with the packet drop function $p(x(t))$, where $x(t)$ is the traffic load in a network node at a time t, as follows:

$$\frac{dE(W)}{dt} = \frac{1}{R(E(q))} - \frac{E(W)E(W(t-\tau))}{2R(E(q(t-\tau)))}p(E(x(t-\tau))), \quad (5.22)$$

where τ is the delay required to alert a TCP source that a packet has been dropped by a network node. $R(.)$ is the RTT function, and $q(.)$ is a function representing the number of packets queued along a given network path. More details on this model and on the performance of AQM approach can be found in [4,57].

References

1. Postel, J. 1980. User Datagram Protocol. RFC 768, August.
2. Comer, D. E. 2000. *Internetworking with TCP/IP: Principles, Protocols, and Architecture*, vol. 1, 4th ed., Upper Saddle River, NJ: Prentice Hall.
3. Stevens, R. W. 1994. *TCP/IP Illustrated*, vol. 1. Reading, MA: Addison-Wesley.
4. Hassan, M., and R. Jain. 2004. *High Performance TCP/IP Networking*. Upper Saddle River, NJ: Prentice Hall.
5. Case, J., M. Fedor, M. Schoffstall, and J. Davin. 1990. A Simple Network Management Protocol (SNMP). RFC 1157, May.
6. Hedrick, C. 1988. *Routing Information Protocol*. RFC 1058, June.
7. Malkin, G. 1994. *RIP Version 2: Carrying Additional Information*. RFC 1723, November.
8. Sportack, M. A. 1999. *IP Routing Fundamentals*. Indianapolis: Cisco Press.
9. Schulzrinne, H., S. Casner, R. Frederick, and V. Jacobson. 1996. RTP: A Transport Protocol for Real-Time Applications. RFC 1889, January.
10. Postel, J. 1981. Transmission Control Protocol. RFC 793, September.
11. Berners-Lee, T., R. Fielding, and H. Frystyk. 1996. Hypertext Transfer Protocol-HTTP/1.0. RFC 1945, May.
12. Fielding, R., J. Gettys, J. Mogul, H. Frystyk, L. Masinter, P. Leach, and T. Berners-Lee. 1999. Hypertext Transfer Protocol-HTTP/1.1. RFC 2616, June.
13. Ameigeiras Gutierrez, P. J. 2003. Packet Scheduling And QoS in HSDPA. PhD diss., Aalborg University, October.
14. TSG RAN Working Group 1, no. 21, HSDPA Number of Users. Qualcomm, R1-01-0912, August 2001.
15. Cohen, E., H. Kaplan, and J. Oldham. 1999. Managing TCP Connections Under Persistent HTTP. *Computer Networks* 31:1709–23 (March).

16. Molina, M., P. Castelli, and G. Foddis. 2001. Web Traffic Modeling Exploiting TCP Connections Temporal Clustering through HTML-REDUCE. *IEEE Network* 14, no. 3:46–55 (May).

17. Postel, J. 1980. DOD Standard: Transmission Control Protocol. RFC 761, 16, January.

18. Mogul, J., and S. Deering. 1990. Path MTU Discovery. RFC 1191, November.

19. Stevens, W. 1997. TCP Slow Start, Congestion Avoidance, Fast Retransmit, and Fast Recovery Algorithms. RFC 2001, January.

20. Paxson, V., and M. Allman. 2000. *Computing TCP's Retransmission Timer.* RFC 2988, November.

21. Stevens, W., M. Allman, and V. Paxson. 1999. TCP Congestion Control. RFC 2581, April.

22. Casetti, C., and M. Meo. 2000. A New Approach to Model the Stationary Behavior of TCP Connections. In Proc. of the IEEE Conference on Computer Communications (INFOCOM) 2000, 1:367–75 (March 26–30).

23. Kumar, A. 1998. Comparative Performance Analysis of Versions of TCP in a Local Network with a Lossy Link. *IEEE/ACM Transactions on Networking*, 6, no. 4:485–98 (August).

24. Casetti, C., and M. Meo. 2001. An Analytical Framework for the Performance Evaluation of TCP Reno Connections. *Computer Networks*, 37:669–82.

25. Padhye, J., V. Frnoiu, and D. Towsley. 1999. A Stochastic Model of TCP Reno Congestion Avoidance and Control. Tech. Rep. 99–02, Department of Computer Science, University of Masschusetts at Amherst.

26. Padhye, J., V. Frnoiu, D. Towsley, and J. Kurose. 1998. Modeling TCP Throughput: A Simple Model and Its Empirical Validation. Paper presented at the annual conference of the Special Interest Group on Data Communication (SIGCOMM), Vancouver, August-September.

27. Padhye, J., V. Frnoiu, D. Towsley, and J. Kurose. 2000. Modeling TCP Throughput: A Simple Model and Its Empirical Validation. *IEEE/ACM Transactions on Networking* 8, no. 2:133–45 (April).

28. Wierman, A., T. Osogami, and J. Olsen. 2003. A Unified Framework for Modeling TCP-Vegas, TCP-SACK, and TCP-Reno. Technical Report CMU-CS- 03-133, Carnegie Mellon University.

29. Jitendra Padhye Web page. "TCP modeling" http://www.icir.org/padhye/tcp-model.html.

30. Brakmo, L., S. O'Malley, and L. Peterson. 1994. TCP Vegas: New Techniques for Congestion Detection and Avoidance. Paper presented at the annual conference of the Special Interest Group on Data Communication (SIGCOMM), August, 24–35.

31. Brakmo, L., and L. Peterson. 1995. TCP Vegas: End to End Congestion Avoidance on a Global Internet. *IEEE Journal on Selected Areas in Communication*, 13, no. 8:1465–80 (October).

32. Mo, J., R. La, V. Anantharam, and J. Walrand. 1999. Analysis and Comparison of TCP Reno and Vegas. Paper presented at the IEEE Conference on Computer Communications (INFOCOM), New York, March, 1556–63.

33. Floyd, S., M. Handley, J. Padhye, and J. Widmer. 2000. Equation-Based Congestion Control for Unicast Applications. Paper presented at the

annual conference of the Special Interest Group on Data Communication (SIGCOMM), Stockholm, August-September.

34. Jacobs, S., and A. Eleftheriadis. 1996. Providing Video Services over Networks without Quality of Service Guarantees. In *World Wide Web Consortium Workshop on Real-Time Multimedia and the Web* (RTMW 96) INRIA Sophia Antipolis, France, October.

35. Ozdemir, V., and I. Rhee. 1999. TCP Emulation at the Receivers (TEAR). Presentation at the IRTF RM meeting, Washington, DC. November.

36. Padhye, J., J. Kurose, D. Towsley, and R. Koodli. 1999. TCP-Friendly Rate Adjustment Protocol for Continuous Media Flows over Best Effort Networks. Paper presented at the Workshop on Network and Operating System Support for Digital Audio and Video (NOSSDAV). Basking Ridge, NJ, June.

37. Rejaie, R., M. Handley, and D. Estrin. 1999. An End-to-End Rate-Based Congestion Control Mechanism for Realtime Streams in the Internet. In Proc. of the IEEE Conference on Computer Communications (INFOCOM), 3:1337–45 (March).

38. Sisalem, D., and H. Schulzrinne. 1998. The Loss-Delay Based Adjustment Algorithm: A TCP-Friendly Adaption Scheme. Paper presented at the Workshop on Network and Operating System Support for Digital Audio and Video (NOSSDAV). Cambridge, UK, July.

39. Wai-Tian, T., and A. Zakhor. 1999. Real-Time Internet Video Using Error Resilient Scalable Compression and TCP-Friendly Transport Protocol. IEEE/ACM Trans. on Multimedia, 1, no. 2:172–186 (June).

40. Turletti, T., S. Pariais, and J. Bolot. 1997. Experiments with a Layered Transmission Scheme over the Internet. Tech. Rep. RR-3296, INRIA, France.

41. Viaisano, L., L. Rizzo, and J. Crowcroft. 1998. TCP-Like Congestion Control for Layered Multicast Data Transfer. In Proc. of the 7th IEEE Conference on Computer Communications (INFOCOM), 3:996–1003 (March-April).

42. TCP-Friendly Web Site: http://www.psc.edu/networking/tcp-friendly.html.

43. Mathis, M., J. Semke, J. Mahdavi, and T. Ott. 1997. The Macroscopic Behavior of the TCP Congestion Avoidance Algorithm. *ACM Computer Communication Review*, 27, no. 3:67–82 (July).

44. Ott, T., J. Kemperman, and M. Mathis. The stationary behavior of ideal TCP congestion avoidance. Unpublished manuscript, ftp://ftp.bellcore.com/pub/tjo/TCPwindow.ps.

45. Balakrishnan, H., M. Stemm, S. Seshan, and R. H. Katz. 1997. Analyzing Stability in Wide-Area Network Performance. Paper presented at the ACM SIGMETRICS. International Conference on Measurement and Modeling of Computer Systems, June, Seattle, WA.

46. Claffy, K., G. Miller, and K. Thompson. 1998. The Nature of the Beast: Recent Traffic Measurements from an Internet Backbone. Paper presented at the conference of the Internet Society INET 98. Geneva, Switzerland, July.

47. Cunha, C. R., A. Bestavros, and M. E. Crovella. 1995. Characteristics of WWW Client-Based Traces. Technical Report BU-CS-95-010, Boston University, July.

48. Gribble, S. D., and E. A. Brewer. 1997. System Design Issues for Internet Middleware Services: Deductions from a Large Client Trace.

Paper presented at the USENIX Symposium on Internet Technologies and Systems. (USITS 97), Monterey, California, December.

49. Mah, B.A., 1997. An Empirical Model of HTTP Network Traffic. Paper presented at IEEE Conference on Computer Communications (INFOCOM) 97, Kobe, Japan (April).

50. Thompson, K., G. J. Miller, and R. Wilder. 1997. Wide-area Internet Traffic Patterns and Characteristics. *IEEE Network* 11, no. 6: 10–23 (November).

51. Heidemann, J., K. Obraczka, and J. Touch. 1997. Modeling the Performance. of HTTP over Several Transport Protocols. *IEEE/ACM Transactions on Networking* 5, no: 5: 616–30 (October).

52. Mahdavi, J. 1997. TCP Performance Tuning. http://w.psc.edu/networking/tcptune/slides, April.

53. Partridge, C., and T. J. Shepard. 1997. TCP/IP Performance over Satellite Links. IEEE Conference on Computer Communications (INFOCOM) 11, no. 5:44–9 (September–October).

54. Cardwell, N., S. Savage, and T. Anderson. 2000. Modeling TCP Latency. In Proc. of IEEE Conference on Computer Communications (INFOCOM) 2000 3:1742–51 (March 26–30).

55. Altman, E., K. Avrachenkov, and C. Barakat. 2000. A Stochastic Model of TCP/IP with Stationary Random Losses. Paper presented at the annual conference of the Special Interest Group on Data Communication (SIGCOMM), Stockholm, Sweden, August.

56. Kelly, F. P., A. K. Maullo, and D. K. H. Tan. 1998. Rate Control for Communication Networks: Shadow Prices, Proportional Fairness, and Stability. *Journal of the Operational Research Society* 49, no. 3:237–52 (March).

57. Misra, V., W. Gong, and D. Towsley. 2000. Fluid-Based Analysis of a Network of AQM Routers Supporting TCP Flows with an Application to RED. Paper presented at the annual conference of the Special Interest Group on Data Communication (SIGCOMM), August.

Chapter 6

TCP over Wireless Systems: Problems and Enhancements

The applications and services built over the TCP/IP protocol stack today represent a large share of the overall traffic volume in the Internet and wireless networks. The TCP control mechanisms originally designed for high bandwidth, short delays, and congestion-limited networks are in fact not suitable for wireless systems. Wireless networks are characterized by high losses due to radio propagation impairments, higher delays, and very scarce bandwidth. Small-scale degradations over the air interface, such as fast fading, induce fluctuations, and losses over the air interface that are mistakenly taken as congestion over the fixed networks by TCP. This occurs despite radio link control mechanisms that typically use retransmissions to achieve error-free communications over the air interface. Radio retransmissions cause delays that are larger compared to TCP timescales, resulting in degradation of end-to-end throughput between distant hosts. In fact, TCP misinterprets errors over wireless links as congestions and reacts by retransmitting TCP segments and by reducing the congestion window and in fine the overall application throughput.

To cope with the wireless link characterizations and to achieve better performance of TCP over wireless networks, several solutions and TCP enhancements have been proposed in the literature. These solutions are suggested for integration in different layers, varying from link layer to transport layer solutions. This chapter provides an overview of these enhancements along with the advantages and drawbacks of each TCP variant.

6.1 Wireless Environment Factors

Before describing the solutions proposed to enhance the TCP performance over wireless links, this section presents a brief overview of the basic factors that affect the TCP characteristics in wireless systems.

6.1.1 Limited Bandwidth and Long RTT

The majority of wireless systems provide low data-transmission rates to users (e.g., wireless local area network [WLAN], UMTS). The limited bandwidth is one the of major factors along with channel impairments that degrade the TCP performance over wireless links.

The Internet and other wired data networks provide higher bandwidth than wireless links: 100 Mbps are common on wired LANs. The difference in bandwidths between Internet and wireless networks affects the behavior of the *last-hop router* [1–4], which connects the wireless network to the other data networks (e.g., GGSN in UMTS). The last-hop router receives more TCP segments than it can route through the wireless network. This generates excessive delays due to segments queueing in the router buffer. These delays increase the RTTs of the TCP connections and inflate the calculated RTO. This limits the increase of the TCP congestion window size, resulting in a limited TCP throughput. In addition, if congestion occurs the fast-recovery phase (in the case of triple duplicate) and the slow-start phase (in the case of timeout) become even more harmful.

The segments queueing in the last-hop router can result in buffer overflow, can prevent packets from dropping at the router, and can cause congestion to increase. This point has been studied in the literature [5,6] to estimate the optimal buffer capacity of the last-hop router. In UMTS (respectively HSDPA), the RLC buffer in the RNC (respectively the MAC-hs buffer in the node B) affects similarly the TCP performance as the last-hop router.

The buffer overflow of the last-hop router explains the *slow-start overshooting* [1–4] behavior observed essentially in the studies of TCP over slow wireless networks (e.g., general packet radio service [GPRS], UMTS). In fact, at the beginning of the TCP connection, the slow-start threshold is set to an arbitrary value that is typically high. The congestion window size (i.e., the number of sent packets) is allowed to grow until a packet loss is detected. Note that the TCP sender detects a congestion one RTT after the congestion took place in the network; the RTT is high due to the low transmission rate over wireless links and the buffering delays in the last-hop router. During this time, the TCP sender, having a window size larger than the network capacity, continues the packet transmission, overloads the network, and makes the congestion more severe. As the congestion occurs, the slow-start threshold progressively gets assigned a proper value.

The congestion-avoidance phase, taking over the window evolution at the end of the slow start, is charged to slow down the sending rate according to the network capacity so that the following overflows are less severe and usually cause a single retransmission.

6.1.2   High Loss Rate

The major cause of packet losses over wireless links is the high level of errors that occur during a transmission. These losses can generate triple-duplicate acknowledgments or timeouts. The triple duplicate occurs when the destination receives an erroneous segment and correctly receives the following segments, which have higher sequence numbers. In this case, TCP acknowledges the error-free segments, which leads to a fast retransmission and fast recovery mechanism in the TCP connection. In addition, low bandwidth, high delay, and losses can generate an absence of feedback control signals from the receiver to the transmitter and consequently can lead to a timeout in the TCP connection.

To deal with the problem of losses and errors, the majority of wireless systems implement a retransmission protocol called ARQ (described in Chapters 3 and 4). The ARQ protocol ensures error-free reception of packets at the receiver over the air interface. It does, however, mislead the TCP protocol that generates unnecessary packet retransmissions through its congestion mechanisms. The lower layer, especially the MAC RLC layer, can receive an erroneous packet and can correctly receive the following packets. The ARQ protocol is in charge of retransmitting the erroneous packet until it is received correctly. In this case, two phenomena can happen. First, the lower layer may deliver the received packets correctly and in sequence to the upper layer. In this case, the TCP layer receives in sequence the TCP segments, but the ARQ protocol causes a delay in receiving segments at the TCP layer. This leads to a delayed acknowledgment on the feedback signals and may generate a TCP timeout in the TCP connection. Note that this scenario happens frequently in UMTS system, where the packets are delivered to the upper layer in sequence. Second, the lower layer may deliver the data correctly but out of sequence. In this case, the TCP layer may receive segments having sequence numbers beyond the expected one. This may generate a triple duplicate or a timeout, according to the delay of receiving the expected segment by the TCP layer.

The randomness and the bursty aspects of the errors over wireless links make the effect of losses more harmful. In fact, the reception of successive error-free packets (i.e., without ARQ delays) reduces the RTO computed from RTT estimations. When the errors occur in bursts, successive erroneous packets are received. This generates excessive delays and severe timeouts, which result in TCP throughput degradations.

Finally, it important to note that spurious timeouts, generated by random errors over wireless links, may result in *retransmission ambiguity* [7]. This means that after the retransmission of a segment interpreted as dropped, the TCP sender does not know whether an acknowledgment arrives for the retransmitted segment or for the first one.

6.1.3   Mobility

The degradation of TCP performance in wireless networks can also be caused in part by the mobility of the users while in communication.

In ad hoc networks, the mobile nodes move randomly, causing frequent topology changes. This results in packet losses and frequent route discovery algorithms initiation, which significantly reduces the TCP throughput.

In cellular networks (e.g., GPRS, UMTS), the user mobility leads to hand-off during the communication. During the handoff process, the necessary information has to be transferred form the previous base station to the new base station. According to the technology used, such as UMTS or HSDPA, the handoff would result in excessive delays or disconnection. The handoff can be intratechnology where both of the cells are covered by the same cellular system, and intertechnology where different technologies are deployed in adjacent cells, such as handoff between UMTS and HSDPA. The intersystem handoff is known to be more harmful, since in some cases the data stored in the RLC or node B buffer is not transferred to the new cell. The lost of data over the air induces drastic degradations of overall TCP throughput.

6.1.4   Asymmetric Links Bandwidth

TCP is a self-clocking protocol that uses the incoming acknowledgments in a direction to estimate the RTT and controls the packet transmission in the opposite direction [8]. In both directions, the delay between the received packets (or the received acknowledgments) depends on the link bandwidth of each direction. To have a normal behavior of TCP, the acknowledgments should maintain the same spacing of the transmitted data in the other direction.

In wireless systems, the downlink and the uplink do not provide the same bandwidth. For example, in UMTS Release 5, the use of HSDPA provides a higher transmission rate on the downlink. On the uplink, the user continues to use the DCH channel, which provides a limited transmission rate. This causes an asymmetry between the downlink and the uplink. In [9, page 1] the network asymmetry is defined as follows: "network exhibits asymmetry with respect to TCP performance, if the throughput achieved is not solely a function of the link and traffic characteristics of the forward (i.e., downlink) direction, but depends significantly on those of the reverse (i.e., uplink) direction as well."

It is important to note that the data segment does not have the same size as the corresponding acknowledgment. In addition to the bandwidth, the size ratio between data segments over the downlink and the correspondent acknowledgments over the uplink, as well as the data size transmitted over the uplink, should be considered to decide if a network exhibits network asymmetry.

The slow-uplink path that carries the acknowledgments can significantly slow down the evolution of the congestion window, which limits the TCP throughput and affects the utilization rate of the downlink. In addition, the acknowledgments sent over the slow uplink may suffer from a dropping if an intermediate router presents a limited buffer space. This leads to less efficient use of the fast-recovery algorithm and degrades TCP performance.

6.2 TCP Performance Enhancements

To enhance TCP performance over wireless systems, many proposals have been developed in the literature, which try to approach one of the two following ideal behaviors: (1) The TCP sender should simply retransmit a packet lost due to transmission errors, without taking any congestion control actions; or (2) the transmission errors should be recovered transparently and efficiently by the network—that is, it should be hidden from the sender. According to their behaviors, the proposed schemes can be classified into three categories [10–13]: link layer solutions, split solutions, and end-to-end solutions. Each type of solution is described to acquire better understanding of the interactions that can occur between radio link layer protocols and TCP.

6.2.1 Link-Layer Solutions

Link-layer solutions aim at making the wireless link layer behave like wired segments with respect to higher-level protocols. The basic idea is that errors over wireless links should be recovered in wireless system without including TCP in the recovery process. In other words, these solutions try to mask or hide the error recovery from TCP. FEC combined with ARQ protocol is used in the majority of wireless systems to provide the reliable service needed by upper layers, such as TCP. As explained in Chapters 1 and 3, FEC results in inefficient use of available bandwidth. ARQ may cause spurious retransmission at the TCP layer, especially when the wireless link suffers from high-level bursty errors. Neither approach is appropriate from the stand point of efficiency or layer interactions.

Additional enhancements must be introduced in the link layer to improve TCP performance. Link-layer solutions can be either TCP aware or unaware. Different assumptions, complexity, and overall system performance are observed in each case.

6.2.1.1 Snoop Protocol

In the majority of the current wireless system, the use of ARQ protocol allows recovery from wireless link errors and provides relatively reliable transfer of packets to the upper layer. However, it was observed that the interaction between ARQ and a reliable transport layer such as TCP may result in poor performance due to spurious retransmissions caused by an incompatible setting of timers at the two layers.

The snooping protocol developed in [14,15] provides a reliable link layer closely coupled with the transport layer so that incompatibility and unnecessary retransmissions are avoided. The main idea is to introduce an agent in the base station, or wireless gateway, that can snoop inside TCP connections to gather the TCP sequence number, to cache the unacknowledged segment in the base station, and to mask the wireless link errors from the TCP sender by crushing the correspondent acknowledgments. Snoop does not consist of a transport layer module but of a TCP-aware link agent that monitors TCP segments in either direction. For this, two procedures—snoop.data for data segments and snoop.ack for acknowledgments—are implemented in the snoop agent at the base station. These two modules work jointly, as follows.

The base station monitors all TCP segments received from the wired host, or sender, and maintains a cache of new unacknowledged packets before forwarding them to the mobile host. The snoop agent decides if a packet is new or not by gathering the sequence number inside the segment header. When an out-of-sequence packet passes through the base station, the packet is marked as retransmitted by the sender to facilitate the behavior of the snoop.ack module once it receives a duplicate acknowledgment of this segment. At the same time, the snoop acknowledgment module keeps track of the acknowledgment transmitted from the mobile host. When a triple duplicate acknowledgment sent by the mobile host corresponds to a packet cached in the base station, the snoop.ack deduces that it is due to wireless link errors. So it retransmits the correspondent packet and suppresses the duplicate acknowledgment. Note that packet losses can also be detected by local timeout since the snoop.ack module maintains an update estimate of the RTT. Further details on the snoop.data and snoop.ack modules can be found in [14].

The main advantage of this scheme is that it maintains the end-to-end semantics of the TCP connection. Another advantage is that it can perform efficiently during a handoff process since cached packets will be transmitted to the mobile as soon as it can receive them. In addition, the snoop relies on TCP acknowledgments to detect whether a packet is received or not, which results in small delays in the case of downlink data transmission—when the TCP receiver, or the mobile host, is near the base station. On the other hand, in the case of uplink data transfer—when the TCP receiver is a remote wired host—TCP acknowledgments are returned too late for an efficient recovery of the lost segments. Consequently, snoop is appropriate for downlink data transfer.

The main disadvantage of snoop is that it requires heavy storage and processing to cache TCP segments. In the case of cellular networks, especially macrocell, the snoop is impractical due to the high number of users in each cell. In addition, a snoop agent will be implemented in the RNC, for example, in UMTS, and not in the base station, which increases the processing complexity and the storage capacity since each RNC can control more than one base station. The presence of SRNC and DRNC to control the base station increases even more the amount of wasted resources inside the UTRAN and the processing complexity, especially when a mobile moves from a cell served by a given RNC to another controlled by another RNC. Consequently, this scheme is adequate for Wireless Local Area Networks (WLANs) rather than for cellular networks [10,11].

Finally, another disadvantage is that the wireless gateway needs to snoop inside TCP connections, which does not work when the packets are encrypted (e.g., IPSec traffic).

6.2.1.2 Transport Unaware Link Improvement Protocol (TULIP)

To recover from wireless link errors and to prevent spurious triple duplicate, Transport Unaware Link Improvement Protocol (TULIP) was proposed in [16]. TULIP is a link-layer protocol that does not require any TCP modifications. In addition, TULIP is TCP unaware, which means that it does not maintain any TCP state and does not require any knowledge about TCP sessions status. More generally, it is transport unaware since it does not know anything about the transport layer; that is, it can be used for TCP or UDP. On the other hand, TULIP is service aware, which means it provides reliable link-layer services for applications relying on TCP—since these application services need this reliability to achieve their QoS—and nonreliable link-layer services for UDP traffic.

The reliability of the link layer is achieved by retransmitting the erroneous packets at this layer without regarding what happens on the TCP layer. The flow control across the link layer is maintained by a sliding-window mechanism. To avoid spurious triple-duplicate acknowledgments and unnecessary congestion control triggers, the link layer delivers the received data in sequence to the upper layer. Note that TULIP does not ensure any reliable service to TCP acknowledgments since "subsequent cumulative ACKs supersede the information in the lost ACK" [16, page 2]. Lost packets are detected by the sender via "a bit vector returned by the receiver as a part of every ACK packet" [16, page 2].

This scheme presents several advantages. First, it does not require any TCP modifications and can deal with any transport layer. Second, it is very simple to implement and does not require any heavy storage or processing to maintain a TCP state like snoop does. Third, it ensures a local reliability by maintaining local recovery of the packet losses at the wireless link without waiting for TCP acknowledgments. Fourth, it can work with any transport or network protocols (e.g., IPv4, IPv6, TCP SACK) even if the transport packets are encrypted, like IPSec. According to [16], this scheme performs well over WLANs and significantly reduces the end-to-end packet delay more than other link-layer solutions.

Note that TULIP resembles the ARQ protocol implemented in UMTS where the RLC layer can use acknowledge and unacknowledge modes and where the RLC SDUs are delivered to the upper layer in sequence (see Chapter 3). Studies conducted on TCP performance in UMTS have shown that the use of schemes such as ARQ and TULIP does not solve the problem of competing retransmissions due to spurious timeouts generated by an incompatible setting of timers at the two layers (see Section 6.1). Moreover, when a packet is erroneous, delaying the subsequent corrected received packets at the link layer may result in more spurious timeouts. This frequently occurs in wireless systems, since packet losses are caused by random burst errors that suddenly span several packets. This aspect was not considered in the performance study in [16], where the conducted simulation relies on two simple loss models: uniform and Markov. The randomness of the burst losses is the main cause of spurious timeouts. Consequently, more research is needed in this direction to improve the TCP performance of TCP over reliable link layer while maintaining the end-to-end semantic of TCP. In [18–20], it was shown that coupling ARQ with enhanced scheduling algorithms at the link layer improves the TCP performance and the system efficiency.

6.2.1.3 Delayed Duplicate Acknowledgments

This is a TCP-unaware scheme [17] that tries to imitate the behavior of the snooping protocol but makes modifications at the receiver rather than at the base station. Therefore, it is preferred over snoop when encryption is used; the intermediate node, such as the base station, does not have to look at the TCP header.

When out-of-sequence packets are received at the TCP receiver (e.g., mobile host, user equipment), the receiver sends duplicate acknowledgments for the first two out-of-order packets. The third and subsequent duplicate acknowledgments are delayed for a duration d. Indeed, this scheme assumes that the out-of-sequence is generated by packet losses over wireless links, and a reliable link-layer protocol such as ARQ is used. Therefore, the erroneous packets will finally be received correctly after a certain delay. Consequently, delaying the third duplicate acknowledgment gives the receiver more time to receive the in-sequence packet and to prevent a retransmission at the TCP layer. If during the delay d the in-sequence segment is not received, the destination releases the deferred duplicated acknowledgments to trigger a retransmission.

Note that determining an appropriate value of d is mandatory to improve the performance of this scheme. A large value of d delays the duplicate acknowledgments and results in a long wasted time, especially when the loss is due to congestion, whereas a small value of d does not give the link layer the necessary time to recover from errors. This scheme considers that the packet losses are due to wireless link errors and not to congestion. This can be considered as the main disadvantage of this technique.

Finally, it is important to note that in UMTS and HSDPA system the packets at the link layer are delivered to the upper layer in sequence (explained in detail in Chapters 3 and 4). The use of the delayed duplicate-acknowledgment technique is not useful since the lower layers (RLC and MAC-hs) perform the same role.

6.2.1.4 Scheduling over Reliable Shared Channel

Before describing this scheme, a brief definition of *dedicated link layer* and *shared link layer* is needed. In a dedicated link layer a fixed fraction of available radio resources (i.e., available bandwidth) is reserved to each user, such as a DCH channel in UMTS where a dedicated channel corresponds to a spreading code allocated to a given user. In this case, the wireless resource is still assigned to the user even during inactivite periods. The objective of a shared link layer is to use radio bearers more efficiently by sharing physical channels dynamically between users, as is done over the HS-DSCH in HSDPA. This type of link layer is more appropriate for packet-based operation, especially for bursty traffic since the channel can be allocated to other users during inactive periods.

The time-shared nature of the link layer in wireless systems combined with scheduling to achieve statistical multiplexing of users or applications over radio bearers can result in considerable TCP performance improvement. This is due not only to the possibility of sharing channels among users to cope with the bursty nature of NRT traffic but also to the added capability of harnessing the dynamic congestion window swing caused either by congestion or wireless losses. The link layer is assumed to be reliable in this analysis; that is, ARQ protocol is used (e.g., HSDPA) since this is almost always the case in wireless systems today.

In the case of the dedicated link layer, the shrinking of the congestion window due to wireless errors results in TCP throughput degradation. Since the radio resource bandwidth allocated to each user is fixed, the TCP throughput degradation induces a system capacity degradation, or, more exactly a system throughput degradation. In the case of the shared link layer, the system throughput is not affected so much by the TCP throughput reduction. The reduction of the congestion window results in less TCP packets delivered to the wireless network and therefore in more inactivity periods. The time-shared nature of the link layer copes with this situation by allocating the channel to other users during these periods of inactivity. The degradation of system throughput is due essentially to the retransmission of erroneous packets. This aspect was studied by the authors in [18–20]. It was shown that the cell-capacity reduction due to packet retransmission does not exceed 5 percent (see Chapter 7). More flexibility can be envisaged to better manage wireless resource sharing to limit the TCP throughput degradation. This can be performed by enhanced scheduling schemes that couple TCP traffic characteristics with short-term radio-channel variations. In HSDPA, the scheduling algorithms presented in Chapter 4 should be enhanced to deal with the dynamic window size of TCP traffic. In [21], enhanced proportional fair has been used to achieve better fairness between TCP connections (a similar approach has been proposed independently in [20]). The basic idea, as stated earlier, is that any reduction of the TCP sender window size releases more room on wireless shared channels for other connections. This increases the radio-transmission rate of other connections when the window size of the first connection is reduced so that the packets going through the base station are served more quickly, including the potential retransmissions of erroneous packets. This improves the performance of the TCP connections during the congestion-window reduction of a given connection. Since network congestion and wireless errors are random, each connection profits dynamically from random window swings of other connections, which limits the TCP throughput degradation. The key in this solution is the appropriate, or best, use of the scheduler to limit packet delays in the base station buffer to use the window oscillation and to achieve better fairness between connections. The refined proportional fair proposed in [21] addresses these scheduling requirements by allocating the channel to the user having the maximum r/S where r is the instantaneous radio transmission rate and S is the mean rate achieved at the TCP layer. A description of this solution, including the derivation and some results, is provided in Chapter 7. Additional details are presented in [18–21].

More recently, the impact of the wireless link scheduler on the TCP system performance was investigated in [22,23]. The authors suggested an interesting approach to improve the long-lived TCP performance while reducing the latency of short TCP flows. These improvements are shown to provide efficient wireless-channel utilization. This idea is to simultaneously use a network-based solution called window regulator and a proportional fair scheduler with rate priority (PF-RP). The window regulator algorithm conveys the instantaneous wireless-channel conditions to the TCP source by using the receiver window field in the acknowledgement packets. This algorithm uses an ACK buffer to absorb the channel variations. The window regulator leads to increased TCP performance for any given buffer size. The PF-RP scheduler, based on the simultaneous use of two schedulers PF and PF with strict priority (PF-SP), differentiates short flows from long flows and assigns different priorities to these flows. This scheduler achieves a trade-off between fairness among users and system throughput maximization. The scheduler also minimizes latency for short flows. The combination of the window regulator and the PF-RP scheduler is shown to improve the performance of TCP Sack by up to 100 percent over a simple drop-tail algorithm for small buffer sizes at the congested router.

In [24], the authors proposed an opportunistic scheduler that reduces the packets delay in the node B buffer, achieves better fairness between users, and improves TCP performance without losing cell capacity. This scheduler, which relies on the same philosophy as the previous one, has the advantage of also being applicable to streaming services.

Link-layer solutions to reduce TCP degradation appear to be the most promising solutions for improving TCP performance over wireless systems that use ARQ and scheduling over the radio link. More research is required to exploit wireless diversity more extensively by taking more advantage of the random short-term channel variations of users to maximize overall system throughput, to reduce TCP performance degradations, and to achieve fairness among TCP flows. To address this challenge enhanced metrics for system efficiency and fairness trade-offs need to be identified. Deeper understanding of the behavior of the wireless systems and TCP are still required.

6.2.1.5   Other Link-Layer Solutions

Many other link-layer solutions have been investigated in the literature. For example, in [25] a new MAC layer protocol called MACAW was proposed and is essentially designed to enhance current MAC layers such as Carrier Sense Multiple Access (CSMA) and FAMA. It proposes to add link-layer acknowledgments and less aggressive backoff policies to reduce the unfairness in the system and to improve the TCP performance.

In [26], the Asymmetric Reliable Mobile Access in Link Layer (AIRMAIL) protocol was proposed. Basically, this protocol is designed for indoor and outdoor wireless networks. It relies on a reliable link layer provided by a combined usage of FEC and local retransmissions like the ARQ protocol, which results in better end-to-end throughput and latency. The FEC implemented in this case incorporates three levels of channel coding that interact adaptively. The coding overhead is changed adaptively to reduce the bandwidth expansion. In addition, this scheme is asymmetric in terms of placing the bulk of the intelligence in the base station and reducing the processing load at the mobile host. Instead of sending each acknowledgment alone, the mobile host combines several acknowledgments into a single event-driven acknowledgment [26]. At the same time, the base station periodically sends a status message. The main drawback of using the event-driven acknowledgment is the delay in receiving acknowledgments and in retransmitting the erroneous packets [12]. This may invoke congestion-control mechanisms and exponential backoff of the timeout timer at the TCP layer.

6.2.2   Split Solutions

To shield the TCP sender from spurious retransmissions caused by wireless errors, several solutions propose to split the TCP connection into two connections at the point where the wired and the wireless networks meet, since these two subnetworks do not have the same characteristics and the same transmission rate. This point is the wireless gateway, or the mobile router, and it changes from a wireless system to another. In cellular networks like GPRS and UMTS for example, it corresponds to the GGSN since the base station is not IP capable, whereas in wireless LANs it corresponds to access points. The connection splitting is handled by a software agent implemented in this wireless gateway. The first connection—from sender to the wireless gateway—still uses TCP, whereas either TCP or other reliable connection-oriented transport protocol can be used between the wireless gateway and the mobile host, or the receiver. Consequently, TCP performance in the first connection is affected only by the congestion in the wired network, and wireless errors are hidden from the sender.

The main disadvantage of these solutions is they violate the end-to-end TCP semantics (e.g., Indirect-TCP, or I-TCP). The acknowledgment of a given packet may be sent to the TCP sender before this packet is received by the mobile host, or mobile receiver. If for any reason, such as gateway crashes or poor channel conditions, this packet is not received by the mobile host, there is no way to retransmit it by the TCP sender. Also, these split solutions require heavy-complex processing and storage capacity, or scalability problem, since the packets should be buffered at the wireless gateway until they are acknowledged. In addition, in case of handoff the state information should be transferred from one gateway to another (e.g., base station to another in WLAN).

6.2.2.1   Indirect-TCP

Indirect-TCP [27,28] is one the earliest TCP proposals to deal with wireless links. This protocol consists of splitting the connection into two connections, the first one between a fixed host, such as a remote server, and a mobile support router located at the beginning of the wireless network, such as GGSN in UMTS or an access point in WiFi. The second connection is established between the mobile host and the MSR so that faster reaction to mobility and wireless errors can be performed. To establish a connection with a remote server, the mobile host requests the MSR to open a TCP connection with the server, or fixed host. Hence, only an image of the mobile host on the MSR is seen by the fixed host. These two connections are completely transparent to each other. In case of handoff, the state information of each connection is transferred from an MSR to another without reestablishing the TCP connection with the fixed host. This hides the wireless effect from the TCP sender and provides better performance of the overall communication between the mobile host and the fixed host than classic TCP implementations, such as Reno or Tahoe.

Although this scheme shields the TCP sender, or fixed host, from the wireless environment effects, the use of TCP in the second subconnection—between the MSR and the mobile host—results in performance problems. The frequent packet losses over wireless link trigger several timeouts and congestion-control mechanisms, causing the TCP sender to stall [10,11]. Also, the violation of the end-to-end TCP semantics and the heavy processing and storage at the MSR, explained herein, represent the other drawbacks of this scheme.

6.2.2.2 Mobile-TCP

During the handoff process, the wireless connection between the mobile host and the base station is stopped for a while to allow the transfer of the user state information between the base stations called a hard handoff. Wireless disconnection is also caused by deep fading conditions over wireless channel. As a result of these disconnections, many TCP segments are delayed or lost, which triggers the congestion-control mechanism and exponential backoff at the TCP sender. This results in low throughput and poor TCP performance.

Mobile-TCP (M-TCP) [29,30] was designed to deal with this situation of frequent wireless disconnection and dynamic wireless link bandwidth while maintaining end-to-end TCP semantics.

The architecture of this scheme adopts a three-level hierarchy with the mobile host at the lowest level, the mobile support station at the cell level, and the supervisor host at the highest level. The mobile support station has the role of communicating with the mobile host and to transfer the packets to the supervisor host, which controls and manages the connections between the mobile host and the fixed host [13,29,36]. The supervisor host controls several mobile support stations, or cells. Using two TCP clients at the supervisor host, the TCP connection is split into two connections: classic TCP connection between the fixed host and the supervisor host; and M-TCP connection between the supervisor host and the mobile host.

To preserve the end-to-end TCP semantics, the supervisor host does not acknowledge a segment received from the fixed host until it receives its acknowledgment from the mobile host. This maintains the timeout timer estimate at the fixed host based on the whole round-trip time. In fact, once a TCP packet is received by the TCP client at the supervisor host, it is delivered to the M-TCP client to be transferred over the wireless link to the mobile host. A timeout timer at the supervisor host is triggered for each segment transmitted over the M-TCP connection. When the acknowledgment of a given TCP segment is received from the mobile host, the supervisor host acknowledges the last TCP byte of the data to the fixed host. The TCP sender then considers that all the data up to the acknowledged byte has been received correctly. If the supervisor host timer expires before receiving an acknowledgment—for example, due to fading, handoff, or low bandwidth—the supervisor host sends an acknowledgment segment for the last byte to the sender. This acknowledgment contains an update of the advertised window size to be set to zero. This freezes the TCP sender, or forces it to be in persist mode, without triggering the congestion-control mechanism and exponential backoff of the timeout timer.

The handoffs at the highest level—the supervisor host level—occur when a mobile moves from a cell controlled by a given supervisor host to another cell controlled by another supervisor host. In other words, the handoffs occur by moving from one supervisor host domain to another and not from one cell to another. Compared to I-TCP, M-TCP allows the mobility of the host to be handled with minimal state information—that is, with lower processing and scalability cost [13].

6.2.2.3 Mobile End Transport Protocol (METP)

Although splitting the TCP connection into two connections attempts to shield the TCP sender from wireless effects, it was observed that the use of the TCP/IP protocol stack for the second connection (i.e., between the base station and the mobile host like in I-TCP) results in performance problems. This can be interpreted, as indicated in [10,11, page 3], by the fact that "TCP is not well turned for the lossy link, so that the TCP sender of the wireless connection often times out, causing the original sender to stall."

Mobile End Transport Protocol (METP) [31] is a new transport protocol that eliminates TCP and IP layers and operates directly over the link layer. By implementing METP on wireless links—that is, on the second connection between the mobile host and the base station—the performance problems illustrated previously can be avoided. The base station, or splitting point, acts as a proxy for a TCP connection providing a conversion of the packets received from the fixed network into METP packets. This results in a reduced header since the transport and IP headers (i.e., source and destination addresses, port and connection related information) are removed.

Compared to TCP Reno, METP enhances the throughput by up to 37 percent [31]. On the other hand, this scheme presents many drawbacks. It violates the end-to-end TCP semantics and increases the complexity of the base station with an increased overhead related to packet processing, such as conversion from TCP/IP to METP and vice versa. In addition, during handoff a large amount of information, including states, sending and receiving windows, and contents of buffers, has to be handed over to the new base station, which results in greatly increased complexity.

6.2.3 End-to-End Solutions

This solution category includes changes to TCP mechanics and involves more cooperation between the sender and the receiver (hence the name *end-to-end*) to separate wireless losses and network congestion. This section presents an overview of the most popular end-to-end TCP enhancement schemes, along with their advantages and drawbacks.

6.2.3.1 TCP SACK

In wireless channels, there is a high likelihood of burst errors that span a few TCP segments in a given window. In this case, the lack of selective acknowledgments in the traditional TCP versions (e.g., Reno) raises a problem. In fact, TCP uses a cumulative acknowledgment scheme that does not provide the TCP sender with sufficient information to recover quickly from multiple losses within a single transmission window. This forces the sender to either wait an RTT to find out about each lost packet or to unnecessarily retransmit segments that have been correctly received. In addition, when multiple segments are dropped successively, TCP may lose its ACK-based clock, which drastically reduces the throughput.

To partially overcome this weakness, selective acknowledgment (SACK) is introduced in TCP [5–7, 34] and is defined as one of the TCP options. (For more about TCP options see for example, [37].) Its activation is negotiated between peer TCP entities during the establishment phase of a TCP connection. If no other TCP options are utilized (e.g., timestamps), the SACK scheme makes it possible to inform the sender about a maximum of four losses in a single transmission window. This allows the sender to recover from multiple packet losses faster and more efficiently. Note that the TCP SACK version uses the same congestion algorithm: When the sender receives three selective duplicate acknowledgments, it retransmits the segment and halves the congestion window size. The number of outstanding segments is stored in the so-called variable *pipe* [33]. During the fast-recovery phase, the sender can transmit new or retransmitted data when *pipe* is less than the congestion window *cwnd*. The value of *pipe* is incremented by one when a segment is transmitted and decremented by one at each duplicate acknowledgment reception. The fast-recovery phase is terminated by an acknowledgment of all the segments that were outstanding at the beginning of the fast recovery.

Finally, even though the improvement gain by using selective acknowledgment is very high compared to Reno, the SACK scheme does not overcome totally the degradation of TCP performance when applied to wireless links. In TCP SACK, the sender still assumes that all packet losses are due to congestion and does not maintain the value of the congestion window size *cwnd* when losses are due to errors. In addition, the SACK scheme does not improve the performance when the sender window size is not sufficiently large [38,39].

6.2.3.2   Forward Acknowledgment

Forward acknowledgment (FACK) [40] makes more intelligent decisions about the data that should be retransmitted. However, it is more or less targeted toward improving the performance of TCP when losses are due to congestion rather than to random losses. This scheme incorporates two additional variables *fack* and *retran.data*. The variable, *fack*, represents the forward-most data acknowledged by selective acknowledgments and is used to trigger fast retransmit more quickly, whereas *retran.data* is used to indicate the amount of outstanding retransmitted data in the network [40,41]. When the sender receives a triple duplicate, TCP FACK regulates the amount of data that should be retransmitted during the fast-recovery phase within one segment of the congestion-window size as follows: *(forward most data sent) − (fack) + (retran.data)*.

6.2.3.3   SMART Retransmissions

Self-Monitoring Analysis and Reporting Technology (SMART), proposed in [42], is an alternate proposal to TCP SACK. Each acknowledgment segment contains the standard cumulative acknowledgment and the sequence number of the packet that caused the acknowledgment. This scheme allows the sender to retransmit lost packets only. In addition, it decouples error and flow control by using two different windows: an error-control window at the receiver for buffering the out-of-order packets; and a flow-control window at the sender for storing unacknowledged packets. This scheme reduces the overhead needed to generate and to transmit acknowledgments at the cost of some "resilience to reordering" [10,11, page 13]. Like TCP SACK, SMART retransmissions do not make any difference between losses due to congestion and those due to wireless link errors.

6.2.3.4   Eiffel

In the case of spurious timeouts, the TCP sender proceeds to a retransmission of the segment interpreted to be dropped. When the sender receives an acknowledgment after the segment retransmission, it cannot decide if that acknowledgment corresponds to the retransmitted segment or to the first one. This phenomenon, explained in Section 6.1.2, is called *retransmission ambiguity*.

The TCP Eiffel was proposed to deal with this situation of retransmission ambiguity [43,44]. For this end, it uses the timestamp option in the TCP header to assign a number to the retransmitted segment [37]. The timestamp value, the current congestion-window size *cwnd*, and the slow-start threshold are stored by the sender. Once an acknowledgment is received, the sender compares the timestamp values of the acknowledgment and the retransmitted segment. If the acknowledgment timestamp value is less than that of the retransmitted segment, the sender concludes that a spurious retransmission has occurred. Therefore, it resumes the transmission with new data and stores the congestion-window size and the slow-start threshold values before the spurious retransmission.

The advantage of this scheme is that it uses one of the TCP options: It does not require a new standardization. On the other hand, use of TCP options increases the size of TCP segments. TCP SACK uses the header options field to indicate up to four lost segments in an RTT. The use of timestamp or other TCP options limits the number of lost segments indicated by the SACK scheme. To overcome this constraint, Eiffel can use a new specific bit in the TCP header to indicate whether an acknowledgment is for a retransmitted segment or for the original one. The disadvantage of this solution is that it requires a new standardization to specify the new bit in the TCP segment header.

Finally, it is important to note that the Eiffel algorithm does not prevent the spurious timeouts, but it detects them after the segments retransmission, which limits the efficient use of the wireless bandwidth.

6.2.3.5 Explicit Congestion Notification

The TCP explicit congestion notification (ECN) [45–47] feature provides a method for an intermediate router to notify the end hosts of impending network congestion. This prevents TCP connections, especially short or delay-sensitive connections, from unnecessary packet drop or retransmission timeouts. Basically, the ECN scheme was not proposed to deal with TCP performance over wireless links. However, the congestion-notification mechanism implemented in this scheme can be extended to help the sender differentiate between losses due to congestion and those due to wireless link errors (for example, see [48–52]).

In the Internet, the use of the active queue management (AQM) gateway, like random early detection (RED), allows detection of a congestion before an overflow of the gateway buffer occurs. The RED gateway drops the probability of the packets according to the average queue space. The ECN scheme enhances the AQM behavior by forcing the gateway to mark the packets instead of dropping them. This avoids unnecessary packet drops and allows the TCP sender to be informed quickly about congestion without waiting for triple duplicates or timeouts. To support the ECN scheme, an extension of the TCP/IP stack should be introduced [45–47]. This extension introduces two new flags in the TCP header reserved field: the ECN echo flag and the congestion window reduced (CWR). In addition, two bits in the header are used: ECT-bit and CE-bit.

In the three-way handshake connection phase, the TCP sender and receiver negotiate their ECN capabilities by setting the CWR and echo flags in the SYN and the ACK segments; however, in the ACK segment only the ECN echo flag is set by the receiver. In case when both of the peer TCP entities are ECN capable, the ECT-bit is set to one in all packets; otherwise it is set to zero.

Once the connection is established, the TCP segments transferred through the network can be dropped or marked by the RED gateways based on the queue status. In fact, if the average queue size, which is an exponentially weighted instantaneous queue size, is between a lower threshold min and a higher threshold max, the probability of the packets are marked using a given function. In this case, the CE-bit is set to one by the RED gateway. The packets are dropped if the average queue length at a given gateway exceeds the higher threshold max. Once the receiver detects a packet with a CE-bit set to one, it sets the ECN echo flag of the subsequent acknowledgment segments until it receives a segment with the CWR flag set. This means that the sender has reacted to the congestion echo by adjusting the slow-start threshold and reduction of the congestion-window size. The sender does not retransmit the marked packets, which prevents the connection from wasting time in the fast-recovery phase.

In [48–52], the ECN-capable TCP was enhanced to deal with wireless-link errors, called wireless ECN (WECN). When a packet is dropped or marked, the receiver sends an acknowledgment with the ECN echo flag set. In case of errors over wireless links, the echo flag is not set in the ACK segments. Therefore, after a triple duplicate acknowledgment or a timeout, the connection does not go into the congestion-control mechanism. However, it can reduce the congestion-window size since "the system is still in bad state." When the wireless conditions become favorable again, the sender recovers quickly [48–52, page 7].

The WECN scheme assumes that TCP entities and all routers are ECN compliant, which is not the case in real networks. This is a big disadvantage of this scheme since it is impractical to reconfigure the entire set of nodes in a network. Therefore, more research work is needed in this direction to develop a more powerful ECN extension.

6.2.3.6 *Explicit Bad State Notification (EBSN)*

The explicit bad state notification (EBSN), proposed in [53,54], tries to avoid spurious timeouts by using an explicit feedback mechanism. A TCP agent is introduced in the base station. This scheme, essentially proposed for TCP downlink data transfers, is not considered as a link-layer solution since it requires a TCP sender support [12].

If the wireless links is in bad state—that is, a packet is received with errors—the ARQ protocol is in charge of retransmitting the packet until it is received correctly. At each packet retransmission, the base station sends an EBSN message to the TCP sender indicating a delay of correct packet transmission. On the receipt of an EBSN message, the TCP transmitter reinitializes the TCP retransmission timer. This provides a reliable link layer that can correctly deliver the packet to the destination without triggering spurious timeouts and unnecessary packets retransmission by the TCP layer.

One of the disadvantages of this scheme is that it requires TCP code modifications at the source to be able to interpret EBSN messages. In addition, the base station must be able to keep track of all the packets going through it to note which packets have been acknowledged or not. On the other hand, no state maintenance is required, and the clock granularity (timer) has little impact on performance [53,54].

6.2.3.7 Explicit Loss Notification

The snooping protocol described in Section 6.2.1 was proposed only for downlink data traffic from a fixed host toward a wireless host. To deal with the degradation of the TCP performance over the wireless uplink— that is, when the source is the wireless host—a scheme called explicit loss notification (ELN) was developed in [55]. Contrary to snoop, this scheme is not a pure link-layer solution since it requires modifications to the transport layers of remote wired hosts.

This scheme consists of implementing an agent at the base station, or wireless gateway, to detect the wireless link losses. The cause of losses— either congestion or wireless errors—is communicated to the sender through a bit in the TCP segment header called ELN. When this ELN bit is set in the TCP acknowledgment segment header, the TCP sender deduces that the packet loss is due to errors over the wireless link. Unlike snoop, the base station in this case keeps track of the TCP sequence numbers without caching the TCP segments. When a duplicate acknowledgment is detected by the base station, the sequence number of the lost segment is extracted from the ACK. The base station verifies if it has received correctly this segment. If it is not the case, the base station sets the ELN bit in the TCP header and forwards the duplicate acknowledgment to the wireless host. The TCP sender reacts to this ACK with ELN bit set by retransmitting the lost packet without triggering any congestion-control mechanism. Consequently, this scheme detects if the packet losses are due to congestion or to wireless link errors and prevents unnecessary reduction in the congestion window.

Although this scheme can detect the exact cause of packet loss, it introduces unnecessary delay for retransmitted packets. Even if the base station detects a packet loss over the wireless link, it should wait for a duplicate acknowledgment from the TCP receiver, or the remote wired host, to ask the wireless host for a retransmission [41].

6.2.3.8 TCP over Wireless Using ICMP Control Messages

The Internet Control Message Protocol (ICMP), proposed in [56], is similar to the EBSN scheme. It allows the sender to distinguish whether losses are likely due to congestion or wireless errors, thereby avoiding the sender from needlessly cutting down its congestion window. If the first attempt of a packet transmission over wireless link fails, an ICMP control message (see [57,58]) called ICMP-DIFFER and containing TCP and IP headers is sent to the TCP sender. Therefore, the sender receives surely either an acknowledgement or an ICMP-DIFFER within one RTT after the packet transmission. At the receipt of the ICMP-DIFFER message, the sender re-sets its retransmission timer according to the current RTT estimate without changing its congestion-window size and slow-start threshold. The timer's postponement of a RTO gives the base station sufficient time to locally retransmit the erroneous packet [56]. In the case of successive failures of the packet transmission, another ICMP message, called ICMP-RETRANSMIT, is sent to the TCP source. At the same time, the TCP receiver generates duplicate acknowledgments since other subsequent packets have been received. At the receipt of ICMP-RETRANSMIT and the first duplicate ACK—concerning the segment indicated by the ICMP message—the sender goes into fast-recovery phase and retransmits the erroneous packet. Once a new acknowledgment is received, the recovery phase is ended and the congestion-window size is set to the value before IMCP-RETRANSMIT was generated (i.e., *cwnd* is not halved like in conventional TCP). If neither ICMP-DIFFER nor ICMP-RETRANSMIT are sent to the sender, the TCP connection follows conventional TCP algorithms once a triple duplicate ACKs or a retransmission timeout occur. It is better to generate ICMP messages at the base station rather than at the TCP receiver, since wireless errors may span all the corrupted packets including the TCP and IP header. On the other hand, this solution requires an intelligent and sophisticated base station capable of keeping track of all the packets going through it and to note whether a packet has been acknowledged or not.

6.2.3.9 Noncongestion Packet Loss Detection (NCPLD)

Noncongestion packet loss detection (NCPLD) is an algorithm proposed in [59] to implicitly determine whether a loss is caused by congestion or wireless-link errors. It can be classified into the category of implicit congestion notification schemes. This solution is an end-to-end scheme that requires modifications at the TCP sender only, which facilitates its implementation.

The main idea is to use the concept of the network knee point. The knee point is the point of the throughput load-graph at which the network operates at optimum power [60,61]. Before the knee point, the network is underutilized, and the throughput increases greatly with the network load. In this case, the RTT increases slowly, and it can be assumed that it remains relatively constant. After the knee point, the round trip delay increases highly since the transmitted packets need to be queued at the network routers. NCPLD consists of comparing the current RTT estimate to the RTT of the knee point, called delay threshold. If the estimated RTT is below the delay threshold, the TCP sender considers that the network does not reach yet its knee point and that the loss is due to wireless errors. The packet is then retransmitted without triggering any congestion mechanism. On the other hand, when the RTT is greater than the threshold, the NCPLD scheme is very conservative in this case and considers the loss to be due to congestion. NCPLD complies with the congestion algorithms of standard TCP in this case.

Compared to EBSN and ECN, NCPLD is less accurate since it is based on measurements and assumptions. The knee point is not the congestion point of the network. Even if the estimated RTT is greater than the delay threshold, the loss can be due to wireless errors, especially when these errors occur in burst and span more than one packet in a transmitted window, which delays the received acknowledgments and inflates the RTT estimate.

6.2.3.10 Explicit Transport Error Notification

Explicit transport error notification (ETEN), proposed in [62–64] for error-prone wireless and satellite environments, is not a pure end-to-end scheme since it requires modifications at the intermediate network nodes. ETEN assumes that when a loss occurs over wireless link, the TCP segment header, containing the source port number and the TCP sequence number, is still intact. Therefore, the sender notified by an ETEN retransmits the lost packet without needlessly reducing the congestion-window size. This scheme is similar to ECN described already.

6.2.3.11 Multiple Acknowledgments

Multiple acknowledgments (MA), proposed in [65], is not a pure end-to-end protocol either. However, it is presented in this section since it requires modifications at the TCP sender, meaning the receiver is not modified. This scheme, based on some existing solutions, such as snoop, ELN, and I-TCP, combines three approaches: end-to-end semantics, link retransmission, and ELN [13]. It is applied only in the downlink data transmission—that is, when the wireless host is the destination and the last hop is wireless.

The main idea is to distinguish between wireless errors and congestion using two types of acknowledgments: partial and complete. The partial acknowledgment informs the sender that the packet has been received by the base station, whereas the complete acknowledgment is the normal TCP acknowledgment indicating that the packet has been received correctly by the receiver. In MA, the base station monitors the packets going though it and buffers all unacknowledged packets. A partial acknowledgment is generated by the base station either if a wireless transmission attempt of a packet fails—when the base station timer expires before receiving an acknowledgment from the mobile host or when a negative acknowledgment is received—or if an out-of-sequence packet is received at the base station and if that packet is in the buffer. The sender reacts to the partial acknowledgment by updating its RTT estimate and postponing its retransmission timer according to the estimated RTT. This prevents the TCP connection from triggering unnecessary retransmissions and congestion mechanisms [13]. In the case where neither partial acknowledgment nor complete acknowledgment are received by the sender, the TCP connection complies with the congestion algorithms—slow start, fast recovery, timeout, and window reduction—of standard TCP Reno.

6.2.3.12 Negative Acknowledgments

This scheme uses the options field of the TCP segment header to send a negative acknowledgment (NACK) to the sender when a packet is received in errors [12–66]. Negative acknowledgments indicate to the sender that no congestion has occurred and then there is no need to trigger the congestion control mechanism. The sender reacts to NACK by retransmitting the erroneous segment without changing the congestion window size nor the RTT estimate. The main drawback of this scheme is that it assumes that the TCP header of an erroneous packet that holds the port number and the source address remains intact, which is not the case over wireless links where bursty errors may span not only the entire TCP packet but also several data packets in more than one TCP window [12]. The NACK cannot be transmitted to the sender in this case, or it will transferred to a wrong TCP source.

6.2.3.13 Freeze TCP

In wireless systems, mobility and dynamic environment produce a fade period in the received signal. In certain conditions, such as temporary obstacles that block the signal, it is possible that the received signal goes to a deep fade for a very long period of time and that a temporary disconnection and reconnection of the wireless communication may occur. The deep fade may span several windows and may generate segment losses across more than one window. In addition, disconnection and losses that are long in duration can be generated by the handoff process (as explained in Section 6.1.3). In certain cases, such as handoff between HSDPA and UMTS

Release 99, the data stored in the RLC or node B buffer are not transferred from the original cell to the new cell due to the complexity of the handoff management process in these cases.

During long fade durations or handoff process, it is more efficient to stop the TCP transmission, even if no congestion has occurred, though without changing the congestion window size. For this end, freeze TCP was proposed [67]. In case of handoff or long fade durations, the TCP receiver can predict if the actual wireless conditions (e.g., fading level, errors, duration of handoff according to the wireless techniques used) will cause a temporary disconnection of the wireless link. If a disconnection is predicted, the TCP receiver sets the advertised window size to zero and transmits it to the sender in an acknowledgment segment. Since the TCP sender selects the minimum between the congestion window and the advertised window to be the transmission window, the zero of the advertised window freezes the sender [67]; that is, it forces the sender to be in a persist mode where it stops sending more packets without changing its congestion window and retransmission timeout timer [68]. Once the receiver detects a reconnection, it transmits several acknowledgments to the sender acknowledging the last received packet before disconnection. This process prevents the sender from an exponential backoff and allows the data transmission to resume promptly at the original transmission rate before the disconnection.

The timing to send out the zero window-size acknowledgment is a critical factor that can affect the performance. Acknowledgment that is too late may trigger timeout and the congestion-control mechanism, whereas early acknowledgment causes stop of packets sending earlier than necessary. Experimental results show that the optimal time to send out the zero window-size acknowledgment is exactly one RTT before the disconnection [67].

6.2.3.14   TCP Probing

TCP probing, proposed in [69,70], introduces a new error-detection-and-recovery strategy capable of ruling on the nature of a detected loss. The main idea is to probe the network and to estimate the RTT to deduce the level of congestion and to trigger the responsive recovery mechanism, thereby avoiding needless *cwnd* shrinks and backoffs. When the sender detects a loss via a triple duplicate or a timeout, it goes into a probe cycle, wherein probe segments are exchanged between the sender and the receiver [71]. The probe segment consists of the TCP segment header without any payload, which alleviates the congestion. The probe cycle is terminated when the sender makes two successive RTT estimations. This means that the probe cycle is extended if a probe segment is lost—that is, if the error persists. In other words, the probe cycle duration is adapted to the channel conditions, which is very useful in the case of persistent bursty errors or handoff between cells. In addition, the interruption of data transmission during the probe cycle allows adaption of the TCP transmission rate to the wireless channel conditions, such as capacity or errors [69,70].

Once the probe cycle is terminated, the sender deduces the congestion level using the measured probe RTTs. If the loss is caused by congestion, such as in the case of persistent errors, TCP probing complies with the congestion-control algorithms of standard TCP, such as fast recovery and timeout [71]. Otherwise, the transmission is resumed without cutting down the congestion-window size. TCP probing does not significantly outperform standard TCP versions (e.g., Reno, Tahoe) in the case of transmission with a small congestion window; however, it is more effective in the case of large sending windows since it avoids the slow-start and congestion phase after wireless losses.

6.2.3.15 Wireless TCP

In Wireless Wide Area Networks (WWAN), the link connection suffers from very low and highly variable bandwidth, bursty random packet errors, and large RTT with high variance [71]. This results in highly variable latency experienced by the endpoints, which may trigger spurious timeouts and exponential timer backoffs at the TCP layer. In addition, the acknowledgments from the mobile host to the fixed host get bunched, which skews the RTT estimation and inflates the calculated RTO. In [72,73], it was shown that the RTT varies between 800 ms and 4 sec and that the RTO may reach 32 sec. It also was observed in [72,73] that WWAN experience occasional blackouts ranging from 10 sec to 10 min during the course of a day. These blackouts are generated by deep fades, temporary lack of available channel, or handoff between nonoverlapping cells.

Vis-à-vis the hard conditions of WWAN, regular TCP suffers from heavy performance degradation. Advanced TCP implementations that use changes in RTT to estimate congestion do not perform well in WWAN since they can be confused by large RTT variations [72,73]. Specific TCP enhancements are then required.

In this context, wireless TCP (WTCP) [72,73] was proposed to improve the TCP performance over WWAN by addressing the causes of throughput degradation. This protocol is designed to be deployable, robust, fair, efficient, and reliable [13]. In the literature many TCP enhancement proposals use the same name of wireless TCP (e.g., [74–76]). In this section, the WTCP presented is the one proposed in [72,73] for WWAN.

WTCP is developed using three key schemes. First, WTCP uses a rate-based transmission control rather than the window-based transmission control used in TCP. The transmission rate is estimated at the receiver using the ratio of the interpacket separation at the receiver and the interpacket separation at the source. The sender transmits the current interpacket separation to the receiver with each data packet and receives the updated estimated transmission rate from the receiver. The packet loss or the retransmission timeouts are not used in the rate control. This makes WTCP insensitive to wireless losses—noncongestion losses—large RTT variations, and bunched acknowledgments. As a result, WTCP is able to handle asymmetric channels and to ensure fairness between competing connections having different RTTs [77].

Second, since RTTs are large over WWAN links and data transmissions may be short-lived, WTCP does not go through the slow-start phase upon startup or to recover from blackouts [72,73]. Rather, it incorporates the packet-pair approach [78] to compute the appropriate initial transmission rate. This is performed by sending two back-to-back packets of maximum-size MSS and computing their interpacket delay.

Third, WTCP ensures the reliability by using periodic cumulative and selective acknowledgment rather than retransmission timeouts, because RTT estimates are skewed by bunched acknowledgments. The sender compares the state of unacknowledged packets in the received acknowledgments to what is stored locally with last retransmission to determine whether an unacknowledged packet is lost or is still in transit. When an acknowledgment is not received by the sender at specified intervals, the sender goes into blackout mode. Like probing TCP, WTCP uses probes to elicit the acknowledgments from the receiver in this case [12,69,70].

6.2.3.16   TCP Peach

TCP peach [79,80] is an end-to-end solution that presents some similarity with TCP probing. This scheme is essentially proposed to solve the slow-start problem in satellite networks and can be used to distinguish between congestion and wireless losses. The main idea is to send low-priority segments called *dummy segments* to probe the available capacity of the network. These dummy segments have the same content as normal segments, but they are served with low priority at the network routers; that is, when a congestion occurs these segments are dropped by the routers. Consequently, these segments do not increase the congestion level of the network. This assumes that some priority mechanism needs to be supported at the network routers [79,80].

In TCP peach, dummy segments are generated in the following three phases [79,80]: sudden start, rapid recovery, and congestion avoidance. In the sudden-start phase of a new connection, the transmission of dummy segments is useful to adapt the slow-start growth since the sender does not know any information about the available resource of the network. When a congestion occurs, the sender goes into rapid-recovery phase and sends dummy segments to detect the nature of loss. In the case of congestion, these segments are discarded by the network routers, nonacknowledgments for these segments are received. The recovery phase complies with the congestion algorithm of standard TCP versions (e.g., Reno, Tahoe). When the loss is due to wireless link errors, an acknowledgment of the dummy segment is received; hence, the sender retransmits erroneous packet without reducing its congestion-window size. Finally, the dummy segments can be used in the congestion-avoidance phase to probe the congestion level of the network, thereby avoiding congestion before it occurs.

6.2.3.17 TCP Vegas

TCP Vegas [81,82] approaches the problem of congestion from another perspective than classic TCP versions such as Reno and Tahoe. By adopting a sophisticated bandwidth-estimation scheme, it estimates the level of congestion before it occurs and consequently prevents unnecessary packet drops. In fact, Vegas estimates the backlogged data in the network (i.e., the congestion level) using the difference between the expected flow rate and the actual flow rate every RTT. The expected flow rate, which represents the optimal throughput that the network can accommodate, is determined by *cwnd/baseRTT*, where *cwnd* is the congestion-window size and *baseRTT* is the minimum round-trip time. The actual flow rate is determined by *cwnd/RTT*, where *RTT* is the current round-trip time estimated as the difference between the current time and the recorded timestamp. During the congestion-avoidance phase and using the estimated backlogged data in the network, Vegas decides to linearly increase or decrease the window size to stabilize the congestion level around the optimal point [81,82]:

$$cwnd = \begin{cases} cwnd + 1 & (expected\,rate - actual\,rate)base\,RTT < \alpha \\ cwnd - 1 & (expected\,rate - actual\,rate)base\,RTT > \beta \\ cwnd & \text{otherwise} \end{cases}, \quad (6.1)$$

where α and β are thresholds that represent, respectively, whether the network is under- or overutilized. Therefore, Vegas tries to keep backlogged packets in the network between α and β.

In addition to the congestion-avoidance scheme, Vegas incorporates modified slow-start and retransmission policies. During the slow start, the sender exponentially increases its congestion window every RTT until the actual rate (*cwnd/RTT*) is less than the expected rate (*cwnd/baseRTT*) by a certain value γ. When a packet is lost or delayed, the sender does not necessarily have to wait for three duplicate acknowledgments to retransmit it. If the current RTT is more than the retransmission timeout value, the packet is retransmitted after the reception of one duplicate acknowledgment. This has the advantage of fast retransmitting lost packets when three duplicate acknowledgments are not received, such as when wireless errors span more than one packet in a window.

The main drawback of this scheme is its reliance on the RTT estimate to adjust the congestion-window size. In case of path asymmetry or path rerouting, the estimated RTT is inaccurate and results in skewed estimations of *cwnd* [83]. This may cause a persistent congestion or an underutilization of the network capacity. Finally, it is important to note that Vegas does not make any difference between losses generated by wireless errors and those due to congestion. On the other hand, the approach used to estimate the backlogged packets in the network constitutes the base of other enhanced TCP schemes developed over wireless networks (e.g., TCP Veno).

6.2.3.18   TCP Santa Cruz

TCP Santa Cruz (TCP-SC) [84] makes use of the options field of the TCP header and attempts to decouple the growth of the congestion window from the number of returning acknowledgments. This is gracious in the cases of path asymmetries, wireless links, and dynamic bandwidth systems. TCP-SC uses a congestion detection and reaction schemes similar in spirit to those used in TCP Vegas, but TCP-SC relies on delay estimates along the forward path rather than the RTT. In addition, this scheme detects the congestion at an early stage and prevents undesired congestion-window reduction. In fact, TCP-SC determines whether a congestion exists or is developing. It also determines the direction of the congestion by estimating the relative delay a packet experiences with respect to another in the forward direction, thereby isolating the forward throughput from reverse path events. By estimating the congestion level and direction, TCP-SC tries to achieve a target operating point for the number of packets in the bottleneck without generating congestion. Finally, TCP-SC provides a better loss-recovery mechanism by making better RTT estimates—including the RTT during retransmission and congestion period—retransmitting promptly lost packets, and avoiding retransmissions of correctly received packets when losses span more than one packet in one window [84].

6.2.3.19   TCP Westwood

TCP Westwood (TCPW) [85–87] is a rate-based end-to-end TCP scheme where the sender adjusts the slow-start threshold and the congestion-window size according to the estimated available bandwidth. The sender keeps track of the rate of received acknowledgments (interACK gap) to estimate the eligible rate estimate (ERE) that reflects the available network resource. TCPW detects a loss upon receiving three triple duplicates or after a timeout timer expiration. In this case, the slow-start threshold and the congestion-window size are adjusted as follows [85,86]: In the case of triple duplicate ACKs, the ssthresh is set to *(ERE * RTTmin)/segmentsize* and *cwnd* to the minimum between the current *cwnd* and ssthresh, whereas in the case of timeout *cwnd* is set to one and ssthresh to the maximum between two and *(ERE ∗ RTTmin)/segmentsize*. TCPW exhibits better performance and fairness than conventional TCP versions such as Reno and SACK [85–88].

To deal with heavy loss environment, [89,90] proposes an enhanced version called TCPW with bulk repeat (BR). This scheme includes a loss discrimination algorithm (LDA) to distinguish between congestive losses and wireless losses using a combination of RTT estimation and the difference between expected and achieved rates. When the sender detects losses caused by wireless errors, it enables the use of the BR mechanisms [89]:

■ bulk retransmission used to retransmit all unacknowledged packets in the current congestion window instead of sending only one lost packet. This is useful to recover promptly from bursty errors that span more than one packet in a window.

■ fixed retransmission timeout instead of exponential backoff of the timer when consecutive timeouts occur.

■ intelligent window adjustment allowing the sender to not reduce *cwnd* to ssthresh after a wireless loss.

6.2.3.20   TCP Veno

TCP Veno [91,92] integrates the advantages of both TCP Reno and Vegas and proposes a scheme to distinguish between congestion and noncongestion states. It adjusts dynamically the slow-start threshold according to whether a packet loss is due to a congestion or wireless errors. In addition, it adjusts the window-size evolution linearly during the congestion-avoidance phase. Veno adopts the same methodology as Vegas to estimate the backlogged data in the network. When the number of backlogged packets is below a threshold β, the sender considers that the loss is random, and the connection is said to have evolved into a noncongestive state. In this case, the sender decreases its congestion window by a factor of 1/5. However, when the number of backlogged packets is higher than β, the loss is considered to be congestive, and the connection in this case adopts the Reno standard to recover from losses (i.e., *cwnd* halved and fast recovery). The efficiency and the performance of this scheme depends greatly on the selected value of the threshold β. Experimentally, it was shown that adopting $\beta = 3$ represents a good setting [92].

6.2.3.21   TCP Jersey

TCP Jersey [93] is an end-to-end transport protocol that assumes the use of network routers capable of handling ECN. This protocol is essentially designed for heterogeneous network containing two environments: wired and wireless. The main idea is to adjust the congestion-window size proactively at an optimal rate according to the network condition. For this, it incorporates two schemes: the available bandwidth estimation (ABE) implemented at the TCP sender; and the congestion warning (CW) configuration intro-duced in the routers. In the case of incipient congestion, the CW-configured routers inform the peer TCP entities by marking all packets going through these routers. This allows the sender to distinguish whether a loss is due to congestion or to wireless link errors. The ABE algorithm allows the sender to estimate the available resource in the network. Consequently, the sender adapts its congestion-window size to an optimal value using the informa-tion collected via ABE and CW schemes.

6.2.3.22   TCP Pacing

TCP pacing [94] is a hybrid between rate-based and window-based control transport schemes. It uses the TCP window to determine the data flow, or number of TCP segments, to send through the network and relies on the rates to deduce the transmission instant of each data packet. The idea is to achieve a rate-controlled packet transmission, thereby avoiding bursty traf-fic that can result in packet losses, delays, and lower performance, such as frequent triple duplicates and timeouts. Two pacing methods can be used to achieve rate control. In the first one, the sender spreads the transmission of a window of packets across the entire RTT [94]—that is, at a rate de-fined by the congestion-control algorithm. The second method consists of spreading the packet transmission by delaying the acknowledgments. This second pacing method is less effective since each acknowledgment may generate multiple data packets at the transmitter.

In the literature, many proposals incorporate the pacing scheme in var-ious contexts. In [95], it was used to correct for the compression of ac-knowledgments due to cross traffic. It also can be used to avoid burstiness in asymmetric networks [9] or when acknowledgments are not available to use for clocking such as to avoid a slow start at the beginning of the connection [96,97] or after a loss [98,99]. Concerning the performance of pacing, it is claimed in [100] that this scheme can improve the TCP perfor-mance over long latency links and high-bandwidth satellite environments. It is important to note that pacing does not distinguish whether the losses are due to congestion or to wireless errors, which limits its improvement of TCP performance in wireless systems.

6.2.3.23   TCP Real

TCP real, proposed in [101,102], is an attempt to enhance the real-time capabilities of TCP in heterogeneous wired or wireless data networks. Unlike standard TCP versions where the sender adjusts the congestion-window size, real is a receiver-oriented base control that adapts the sending window size by implementing the wave concept presented in [103–105]. The wave concept can be regarded as a congestion window with fixed size during each RTT and is known by both of the TCP end entities. It consists of sending a side-by-side predetermined number of fixed data segments within one RTT so that the receiver estimates the network congestion level based on the successive segments received—that is, the perceived data rate. According to the congestion level, the receiver adjusts the wave level, or the number of segments the wave contains, and transmits it to the sender in the options field attached to acknowledgment segments [101,102]. The sender reacts to the ACK by adapting its congestion window to the wave level—to the network conditions—which allows the sender to avoid congestion and thereby undesired timeout and recovery phases.

In the case of path asymmetry—that is, when the reverse path is slower than the forward path—TCP real allows decoupling the size of the congestion window from the timeout [71]. Since the asymmetry does not affect the perceived congestion level of the forward path [9], the congestion window is still intact. However, the RTO should be increased due to the delays of the reverse path, retarding the receipt of the acknowledgments.

Potential errors over wireless links do not affect the perceived data rate at the receiver [101,102]. This allows the receiver to distinguish between congestion and wireless errors to recover promptly from wireless losses and to avoid needless congestion-window reduction. When a timeout is triggered, the TCP sender goes into backoff-like standard TCP versions. However, it does not use slow start and adjusts its congestion window quickly to the appropriate wave level when the timeout is due to errors on wireless links.

6.2.3.24 Ad Hoc TCP

Ad Hoc TCP (ATCP) is proposed in [106] to deal with the network partitions problem in the ad hoc network. This scheme is an end-to-end cross-layer solution. In fact, it keeps TCP intact and introduces a new layer between IP and TCP layers called the ATCP layer. In addition, it relies on the use of ECN [45–47] and ICMP messages [56] to indicate, respectively, whether a congestion has occurred or the network is partitioned. In the case of congestion, the ATCP layer, informed by ECN messages, puts the TCP sender into congestion state—that is, the connection goes into the recovery phase. Using the ECN message prevents the sender from waiting for retransmission timeout. In the case of network partition signaled by ICMP messages, ATCP freezes the TCP sender, or forces it into persist mode. Finally, when a loss is generated by wireless transmission errors, ATCP retransmits the erroneous packet for TCP.

By providing an extensive state-of-the-art for TCP over wireless, this chapter reveals most of the currently used approaches to handle interactions between radio link-layer protocols and TCP. Most rely on masking link-layer corrective actions to TCP or on modifying TCP itself. In some cases, however, it is possible to rely exclusively on scheduling mechanisms available in the wireless packet networks themselves to alleviate, if not eliminate, the negative interactions between ARQ over the radio link and TCP. Chapter 7 addresses end-to-end packet transmission for the UMTS HSDPA system. This enhanced version of the UMTS air interface introduces HARQ, AMC, and fast scheduling to enhance data rates and to improve capacity. The presence of scheduling in this system can be used to handle the interactions with TCP without making any changes to TCP itself and without modifying the UMTS HARQ and AMC mechanisms.

References

1. Gurtov, A. 2002. Efficient Transport in 2.5G 3G Wireless Wide Area Networks. PhLic diss. C-2002-42, Department of Computer Science, University of Helsinki, September, 1.
2. Gurtov, A. 2000. TCP Performance in the Presence of Congestion and Corruption Losses. Master's thesis, C-2000-67, Department of Computer Science, University of Helsinki, December.
3. Kuhlberg, P. 2001. Effect of Delays and Packet Drops on TCP-Based Wireless Data Communication. Master's thesis, C-2001-7, Department of Computer Science, University of Helsinki, February.
4. Sarolahti, P. 2001. Performance Analysis of TCP Enhancements for Congested Reliable Wireless Links. Master's thesis, C-2001-8, Department of Computer Science, University of Helsinki, February.
5. Ludwig, R., B. Rathonyi, A. Konrad, K. Oden, and A. Joseph. 1999. Multilayer Tracing of TCP over a Reliable Wireless Link. In Proc. of the ACM SIGMETRICS International Conference on Measurement and Modeling at Computer Systems, Atlanta, 144–154 (May).
6. Kojo, M., K. Raatikainen, M. Liljeberg, J. Kiiskinen, and T. Alanko. 1997. An Efficient Transport Service for Slow Wireless Links. *IEEE Journal on Selected Areas in Communications* 15, no. 7:1337–48 (September).
7. Karn, P., and C. Partridge. 1987. Improving Round-Trip Estimates in Reliable Transport Protocols. In Proc. of ACM Annual Conference of the Special Interest Group on Data Communication (SIGCOMM) 87, 2–7 (August).
8. Jacobson, V. 1988. Congestion Avoidance and Control. In Proc. of ACM Annual Conference of the Special Interest Group on Data Communication (SIGCOMM) 88, Stanford, CA, 314–29 (August).
9. Balakrishnan, H., V. N. Padmanabhan, and R. H. Katz. 1997. The Effect of Asymmetry on TCP Performance. In Proc. of ACM Conference on Mobile Computing and Networking (MOBICOM) 97, Budapest (September).
10. Balakrishnan, H., V. N. Padmanabhan, S. Seshan, and R. H. Katz. 1996. A Comparison of Mechanisms for Improving TCP Performance over Wireless Links. In Proc. of the ACM Annual Conference of the Special Interest Group on Data Communication (SIGCOMM) 96, Stanford, CA, 256–67.
11. Balakrishnan, H., V. Padmanabhan, S. Seshan, and R. H. Katz. 1997. A Comparison of Mechanisms for Improving TCP Performance over Wireless Links. *IEEE/ACM Transactions on Networking* 5, no. 6: 756–69 (December).

12. Fahmy, S., V. Prabhakar, S. R. Avasarala, and O. Younis, 2003. TCP over Wireless Links: Mechanisms and Implications. Technical report CSD-TR-03-004, Purdue University.

13. Natani, A., J. Jakilnki, M. Mohsin, and V. Sharma. 2001. TCP for Wireless Networks. Technical report. University of Texas at Dallas, November.

14. Balakrishnan, H., S. Seshan, and R. H. Katz. 1995. Improving Reliable Transport and Handoff Performance in Cellular Wireless Networks. *ACM Wireless Networks*, 1, no. 4, 469–81.

15. Balakrishnan, H., S. Seshan, E. Amir, and R. H. Katz. 1995. Improving TCP/IP Performance over Wireless Networks. In Proc. of 1st ACM Conference on Mobile Computing and Networking (ACM Conference on Mobile Computing and Netuoring (MOBICOM 95), Berkeley, California, November.

16. Parsa, C., and J. J. Garcia-Luna-Aceves. 1999. TULIP: A Link-Level Protocol for Improving TCP over Wireless Links. In Proc. IEEE Wireless Communications and Networking Conference (WCNC 99), New Orleans, Louisiana, vol. 3 (September).

17. Vaidya, N. H., M. Mehta, C. Perkins, and G. Montenegro. 1997. Delayed Duplicate Acknowledgements: A TCP-Unaware Approach to Improve Performance of TCP over Wireless Links. Technical Report 99-003, Texas A&M University, College Station, February.

18. Assaad, M., and D. Zeghlache. 2005. How to Minimize the TCP Effect in a UMTS-HSDPA System. *Wiley Wireless Communications and Mobile Computing Journal*, 5, no. 4: 473–85 (June).

19. Assaad, M., B. Jouaber, and D. Zeghlache. 2004. Effect of TCP on UMTS/HSDPA System Performance and Capacity. *IEEE Global Telecommunications Conference*, Dallas, Vol. 6, 4104–8 November 29–December 3.

20. Assaad, M., and D. Zeghlache. 2006. Cross-Layer Design in HSDPA System. *IEEE Journal on Selected Areas in Communications*, 24, no. 3:614–25 (March).

21. Klein, T. E., K. K. Leung, and H. Zheng. 2004. Improved TCP Performance in Wireless IP Networks through Enhanced Opportunistic Scheduling Algorithms. IEEE GLOBECOM 04, San Francisco, 5 November 29–December 3, 2744–8.

22. Choon Chan, M., and R. Ramjee. 2004. Improving TCP/IP Performance over Third Generation Wireless Networks, Paper presented at the IEEE Conference on Computer Communication (INFOCOM), Hong Kong, March.

23. Choon Chan, M., and R. Ramjee. 2002. TCP/IP Performance over 3G Wireless Links with Rate and Delay Variation. Paper presented at ACM Conference on Mobile Computing and Networking (MOBICOM)02, Atlanta, GA, September.

24. Assaad, M., and D. Zeghlache. Opportunistic Scheduler for HSDPA System. *IEEE Transactions on Wireless Communications*, forthcoming.

25. Gerla, M., K. Tang, and R. Bagrodia. 1999. TCP Performance in Wireless Multihop Networks. In Proc. of IEEE Workshop on Mobile Computing Systems and Applications (WMCSA), New Orleans, LA, February.

26. Ayanoglu, E., S. Paul, T. F. LaPorta, K. K. Sabnani, and R. D. Gitlin. 1995. AIRMAIL: A Link-Layer Protocol for Wireless Networks. *ACM Wireless Networks,* 1, no. 1:47–60.

27. Bakre, A., and B. R. Badrinath. 1995. Handoff and System Support for Indirect TCP/IP. Paper presented at 2nd the USENIX Symposium on Mobile and Location-Independent Computing. Ann Arbor, MI, April.

28. Badrinath, B. R., and A. Bakre. 1995. I-TCP: Indirect TCP for Mobile Hosts. Paper presented at the 15th International Congerence on Distributed Computing Systems, Vancouver, Canada, May, 136–46.

29. Brown, K., and S. Singh. 1996. M-TCP: TCP for Mobile Cellular Networks. In Proc. of IEEE Conference on Computer Communication (INFOCOM) 96, San Francisco.

30. Brown, K., and S. Singh. 1997. M-TCP: TCP for Mobile Cellular Networks. *ACM Computer Communication Review,* 27, no. 5:19–43 (October).

31. Wang, K. Y., and S. K. Tripathi. 1998. Mobile-End Transport Protocol: An Alternative to TCP/IP over Wireless Links. In Proceedings of IEEE Conference on Computer Communication (INFOCOM) 98, San Francisco, California, vol. 3:1046–53 (March/April).

32. The WAP forum, http://www.wapforum.org, 2001.

33. Mathis, M., J. Mahdavi, S. Floyd, and A. Romanow. 1996. Selective Acknowledgment option, RFC-2018, October.

34. Floyd, S., J. Mahdavi, M. Mathis, and M. Podolsky. 2000. An Extension to the Selective Acknowledgement (SACK) Option for TCP. RFC 2883, July.

35. Mathis, M., and J. Mahdavi. 1996. Forward Acknowledgment: Refining TCP Congestion Control. In Proc. of the ACM Annual Conference of the Special Interest Group on Data Communication (SIGCOMM) 96 (August).

36. Fall, K., and S. Floyd. 1996. Simulation-Based Comparisons of Tahoe, Reno, and SACK TCP. *ACM Computer Communication Review,* 26:5–21 (July).

37. Jacobson, V., R. Braden, and D. Borman. 1992. TCP Extensions for High Performance. RFC-1323, May.

38. Allman, M., S. Floyd, and C. Partridge. 2002. Increasing TCPs Initial Window. RFC 3390, October.

39. Balakrishnan, H., V. Padmanabhan, S. Seshan, M. Stemm, and R. H. Katz. 1998. TCP Behavior of a Busy Internet Server: Analysis and Improvements. In Proc. of IEEE Conference on Computer Communication (INFOCOM) 98, San Francisco, 252–62 (March).

40. Mathis, M., and J. Mahdavi. 1996. Forward Acknowledgement: Refining TCP Congestion Control. In Proc. of the ACM Annual Conference of the Special Interest Group on Data Communication (SIGCOMM), Stanford, CA, 2, no. 4:281–92. (August).

41. Hassan, M., and R. Jain. 2004. *High Performance TCP/IP Networking.* Upper Saddle River, NJ: Prentice Hall.

42. Keshav, S., and S. P. Morgan. 1997. SMART Retransmission: Performance with Random Losses and Overload. In Proc. of IEEE Conference on Computer Communication (INFOCOM)97, Kobe, Japan, 3:1131–8 (April).

43. Ludwig, R., and R. H. Katz. 2000. The Eifel Algorithm: Making TCP Robust against Spurious Re-Transmission. *ACM Computer Communication Review* 30, no. 1, 30–6 (January).

44. Ludwig, R., and M. Meyer. 2002. The Eiffel Detection Algorithm for TCP. Internet Draft, IETF, February.

45. Floyd, S. 1994. TCP and Explicit Congestion Notification. *ACM Computer Communication Review* 24, no. 5:10–23 (October).

46. Ramakrishnan, K., and S. Floyd. 1999. A Proposal to Add Explicit Congestion Notification (ECN) to IP. RFC2481, IETF, January.

47. Ramakrishnan, K., S. Floyd, and D. Black. 2001. The Addition of Explicit Congestion Notification (ECN) to IP. RFC3168, IETF, September.

48. Peng, F., and J. Ma. 2000. An Effective to Improve TCP Performance in Wireless/Mobile Networks. Internet Draft, draft-fpeng-wecn-01.txt, January.

49. Peng, F. 2000. A Proposal to Add Fast Congestion Notification to IP and Improve TCP Performance in Wireless and Mobile networks. Internet draft, draft-fpeng-fcn-01.txt, January.

50. Peng, F., and J. Ma. 2000. A Proposal to Apply ECN into Wireless and Mobile Networks. Internet draft, draft-fpeng-ecn-01.txt, January.

51. Peng, F., S. Cheng, and J. Ma. 2000. An Effective Way to Improve TCP Performance in Wireless/Mobile Networks. In Proc. of IEEE/AFCEA EUROCOM2000, Information Systems for Enhanced Public Safety and Security, Munich, 250–255.

52. Ramani, R., and A. Karandikar. 2000. Explicit Congestion Notification (ECN) in TCP over Wireless Networks. In *Proceedings of IFIP International Conference on Personal Wireless Communications*, Golansk, Poland, 495–99. (September).

53. Bakshi, B. S., P. Krishna, N. H. Vaidya, and D. K. Pradhan. 1996. Improving Performance of TCP over Wireless Networks. Technical Report 96–014, Texas A&M University, College Station.

54. Bakshi, B. S., P. Krishna, N. H. Vaidya, and D. K. Pradhan. 1997. Improving Performance of TCP over Wireless Networks. Paper presented at the 17th International Conference on Distributed Computing Systems, Baltimore, MD, May.

55. Balakrishnan, H., and R. H. Katz. 1998. Explicit Loss Notification and Wireless Web Performance. Paper presented at the IEEE Globecom Internet Mini-Conference, Sydney, Australia, November.

56. Goel, S., and D. Sanghi. 1998. Improving TCP Performance over Wireless Links. *Proceedings of the IEEE International Conference on Global Connectivity in Energy, Computer, Communication and Control*, 2:332–5 (December 17–19).

57. Postel, J. 1981. Internet Control Message Protocol. RFC-792, September.

58. Braden, R. 1989. Requirements for Internet Host Communication Layers. RFC-1122, October.

59. Samaraweera, N. K. G. 1999. Non-congestion Packet Loss Detection for TCP Error Recovery Using Wireless Links. In *Communications* 146, no. 4:222–30 (August).

60. Jain, R. 1989. A Delay-Based Approach for Congestion Avoidance in Interconnected Heterogeneous Computer Networks. *ACM Computer Communication Review* 19, no. 5:56–71.

61. Brakmo, L., and S. Peterson. 1995. TCP Vegas: End to End Congestion Avoidance on a Global Internet. *IEEE Journal on Selected Areas in Communications* 13, no. 8:1465–80 (October).

62. Krishnan, R., M. Allman, C. Partridge, and J. P. G. Sterbenz. 2002. Explicit Transport Error Notification for Error-Prone Wireless and Satellite Networks. BBN Technical Report No. 8333, BBN Technologies, February 7 (revised March 22).

63. Krishnan, R., M. Allman, C. Partridge, J. P. G. Sterbenz, and W. Ivancic. 2002. Explicit Transport Error Notification (ETEN) for Error-Prone Wireless and Satellite Networks—Summary. Paper presented at the Earth Science Technology Conference, Pasadena, CA, June 11–13.

64. Krishnan, R., J. P. G. Sterbenz, W. M. Eddy, C. Partridge, and M. Allman. Explicit Transport Error Notification (ETEN) for Error-Prone Wireless and Satellite Networks, unpublished. See http://www.ir.bbm.com/~Krash/#pubs.

65. Biaz, S., M. Mehta, S. West, and N. H. Vaidya. 1997. TCP over Wireless Networks Using Multiple Acknowledgements. Technical Report 97-001, Texas A&M University, College Station, January.

66. Chan, A., D. H. K. Tsang, and S. Gupta. 1997. TCP (Transmission Control Protocol) over Wireless Links, In Proc. of the 47th IEEE Vehicular Technology Conference 97, 1326–30.

67. Goff, T., J. Moronski, D. S. Phatak, and V. Gupta. 2000. Freeze-TCP: A True End-to-End Enhancement Mechanism for Mobile Environments. In Proc. of IEEE Conference on Computer Communication (INFOCOM 2000), 1537–45.

68. Tian, Y., K. Xu, and N. Ansari. 2005. TCP in Wireless Environments: Problems and Solutions, *IEEE Communications Magazine* 43, no. 3:S27–S32 (March).

69. Tsaoussidis, V., and H. Badr. 1999. TCP-Probing: Towards an Error Control Schema with Energy and Throughput Performance Gains. Technical Report TR4-20-2000, Department of Computer Science, State University New York Stony Brook, April.

70. Tsaoussidis, V., and H. Badr. 2000. TCP-Probing: Towards an Error Control Schema with Energy and Throughput Performance Gains. Paper presented at the 8th IEEE International Conference on Network Protocols, November, 12–21.

71. Tsaoussidis, V., and I. Matta. 2002. Open Issues on TCP for Mobile Computing, *Journal of Wireless Communications and Mobile Computing* 2, no. 1: 3–20 (February).

72. Sinha, P., N. Venkitaraman, R. Sivakumar, and V. Bharghavan. 2002. WTCP: A Reliable Transport Protocol for Wireless Wide-Area Networks. Paper presented at the ACM Conference on Mobile Computing and Netuoring (MOBICOM) 99, Seattle, WA, August.

73. Sinha, P., T. Nandagopal, N. Venkitaraman, R. Sivakumar, and V. Bharghavan. 2002. WTCP: A Reliable Transport Protocol for Wireless

Wide-Area Networks. *ACM/Baltzer Wireless Networks Journal* 8, no. 2–3:301–16 (March–May).

74. Ratnam, K., and I. Matta. 1998. WTCP: An Efficient Mechanism for Improving TCP Performance over Wireless Links. Paper presented at the 3rd IEEE Symposium on Computers and Communications, June–July 30 2, 74–8.

75. Ratnam, K., and I. Matta. 1998. Effect of Local Retransmission at Wireless Access Points on the Round Trip Time Estimation of TCP. Paper presented at the 31st Annual Simulation Symposium, 150–6 (April).

76. Li, Y., and L. Jacob. 2003. Proactive-WTCP: An End-to-End Mechanism to Improve TCP Performance over Wireless Links. Paper presented at the 28th Annual IEEE International Conference on Local Computer Networks, 449–57 (October).

77. Lakshman, T. V., and U. Madhow. 1997. The Performance of TCP/IP for Networks with High Bandwidth Delay Product and Random Loss. IEEE/ACM Trans. On Networking, 5, no. 3:336–50 (June).

78. Keshav, S. 2001. Congestion Control in Computer Networks. PhD diss. University of California at Berkeley, September.

79. Akyildiz, I. F., G. Morabito, and S. Palazzo. 2001. TCP-Peach: A New Congestion Control Scheme for Satellite IP Networks. *IEEE/ACM Trans. on Networking* 9, no. 3:307–21 (June).

80. Akyildiz, I. F., X. Zhang, and J. Fang. 2002. TCP-Peach: Enhancement of TCP-Peach for Satellite IP Networks. *IEEE Communications Letters* 6, no. 7:303–5 (July).

81. Brakmo, L., S. O'Malley, and L. Peterson. 1994. TCP Vegas: New Techniques for Congestion Detection and Avoidance. In Proc. of the ACM Annual Conference of the Special Interest Group on Data Communication (SIGCOMM 94), London, August, 24–35.

82. Brakmo, L., and L. Peterson. 1995. TCP Vegas: End to End Congestion Avoidance on a Global Internet. *IEEE Journal on Selected Areas in Communication* 13, no. 8:1465–80 (October).

83. Mo, J., R. La, V. Anantharam, and J. Walrand. 1999. Analysis and Comparison of TCP Reno and Vegas. Paper presented at the IEEE Conference on Computer Communication (INFOCOM), New York, March, 1556–63.

84. Parsa, C., and J. J. Garcia-Luna-Aceves. 1999. Improving TCP Congestion Control over Internets with Heterogeneous Transmission Media. Paper presented at the 7th Annual IEEE International Conference on Network Protocols, Toronto, Canada, November, 213–21.

85. Mascolo, S., C. Casetti, M. Gerla, S. S. Lee, and M. Sanadidi. 2000. TCP Westwood: Congestion Control with Faster Recovery. Technical Report #200017. Department of Computer Science, University of California at Los Angeles.

86. Casetti, C., M. Gerla, S. Mascolo, M. Y. Sanadidi, and R. Wang. 2001. TCP Westwood: Bandwidth Estimation for Enhanced Transport over Wireless Links. Paper presented at the ACM Conference on Mobile Computing and Networking (MOBICOM), Rome, Italy, July 16–21, 287–97.

87. Gerla, M., M. Y. Sanadidi, R. Wang, A. Zanella, C. Casetti, and S. Mascolo. 2001. TCP Westwood: Congestion Window Control Using Bandwidth Estimation. In Proc. of IEEE Globecom, 3:1698–1702 (November).

88. Zanella, A., G. Procissi, M. Gerla, and M. Y. Sanadidi. 2001. TCP Westwood: Analytic Model and Performance Evaluation. In Proc. of IEEE Globecom, 3:1703–7 (November).

89. Yang, G., R. Wang, F. Wang, M. Y. Sanadidi, and M. Gerla. 2002. TCP Westwood with Bulk Repeat for Heavy Loss Environments. Technical Report #020023. Department of Computer Science, University of California at Los Angeles.

90. Yang, G., R. Wang, M. Y. Sanadidi, and M. Gerla. 2002. Performance of TCPW BR in Next Generation Wireless and Satellite Networks. Technical Report #020025. Department of Computer Science, University of California at Los Angeles.

91. Fu, C. P. 2001. TCP Veno: End-to-End Congestion Control over Heterogeneous Networks. PhD diss., Chinese University of Hong Kong.

92. Fu, C. P., and Soung C. Liew. 2003. TCP Veno: TCP Enhancement for Transmission over Wireless Access Networks. *IEEE Journal On Selected Areas in Communications* 21, no. 2:216–28 (February).

93. Xu, K., Y. Tian, and N. Ansari. 2004. TCP-Jersey for Wireless IP Communications. *IEEE Journal on Selected Areas in Communications* 22, no. 4:747–56 (May).

94. Aggarwal, A., S. Savage, and T. Anderson. 2000. Understanding the Performance of TCP Pacing. In Proc. of IEEE Conference on Computer Communication (INFOCOM), 3:1157–65 (March).

95. Zhang, L., S. Shenker, and D. D. Clark. 1991. Observations on the Dynamics of a Congestion Control Algorithm: The Effects of Two Way Traffic. Paper presented at the ACM Annual Conference of the Special Interest Group on Data Communication (SIGCOMM) 91 Conference on Communications Architectures and Protocols, September 133–47.

96. Aron, M., and P. Druschel. 1998. TCP Improving Startup Dynamics by Adaptive Timers and Congestion Control, Technical Report TR98-318, Rice University, Houston, TX.

97. Padmanabhan, V. N., and R. H. Katz. 1998. TCP Fast Start: A Technique for Speeding Up Web Transfers. Paper presented at the IEEE Globecom 98 Internet Mini-Conference, Sydney, Australia, November.

98. Hoe, J. 1995. Start-Up Dynamics of TCPs Congestion Control and Avoidance Schemes. Master's thesis, MIT, Cambridge, MA, June.

99. Mathis, M., J. Semke, J. Madhavi, and K. Lahey. 1999. The Rate-Halving Algorithm for TCP Congestion Control. Internet Draft, July.

100. Partridge, C. 1998. ACK Spacing for High Bandwidth-Delay Paths with Insufficient Buffering. Intemet Draft, draft-rfced-infopartridge-01.txt, September.

101. Zhang, C., and V. Tsaoussidis. 2001. TCP Real: Improving Real-Time Capabilities of TCP over Heterogeneous Networks. Paper presented at the 11th IEEE/ACM Workshop on Network and Operating System Support for Digital Audio and Video (NOSSDAV), New York.

102. Zhang, C., and V. Tsaoussidis. 2002. TCP-Real: Receiver-Oriented Congestion Control. *Computer Networks* 40, no. 4:477–97 (November).

103. Tsaoussidis, V., H. Badr, and R. Verma. 1999. Wave & Wait Protocol (WWP) An Energy-Saving Transport Protocol for Mobile IP-Devices.

Paper presented at the 7th IEEE International Conference on Network Protocols.

104. Tsaoussidis, V., A. Lahanas, and H. Badr. 2000. The Wave and Wait Protocol: High Throughput and Low Energy Expenditure for Mobile-IP Devices. Paper presented at the 8th IEEE Conference on Networks, Singapore.

105. Tsaoussidis, V., A. Lahanas, and C. Zhang. 2001. The Wave and Probe Communication Mechanisms. *Journal of Supercomputing* 20, no. 2:115–35 (September).

106. Liu, J., and S. Singh. 2001. ATCP: TCP for Mobile Ad Hoc Networks. *IEEE Journal on Selected Areas in Communications* 19, no. 7:1300–15.

Chapter 7

TCP Performance over UMTS-HSDPA System

Third-generation and beyond cellular systems such as UMTS and enhancements like HSDPA are conceived to offer users, in addition to speech, new multimedia services with high-quality image and video with access to private and public data networks such as the Internet. Such advanced wireless techniques often rely on adaptive modulation and coding, retransmission of erroneous radio link blocks, and scheduling algorithms to counterchannel impairments, thereby achieving improved spectral efficiency and system capacity.

The main source of packet loss in wireless systems is the link errors generated by unperfect transmission adaptation to short-term channel variations. Static or fixed link-protection techniques, such as channel coding and interleaving, are not effective in providing link protection and in correcting all errors experienced over the radio link. The use of ARQ to retransmit erroneous packets is mandatory to achieve error-free radio transmission. However, introducing ARQ incurs additional delays in packet delivery due to retransmissions. These delays conflict with TCP control mechanisms that interpret delays in packet delivery over the wireless link as congestion in the fixed and Internet segments. Since the majority of the Internet services relies on TCP, this protocol is expected to handle a large share of the overall amount of nonreal-time data traffic conveyed over wireless systems. The presence of TCP in the end-to-end path between hosts is a fact that must be taken into account when introducing advanced techniques in wireless. Ideally, changes to TCP should also be avoided since it has already been widely deployed over the past decade. TCP Reno is the most implemented version and is extensively used by Internet applications and services. Chapter 6, providing the state-of-the-art on wireless TCP, shed some light already on expected interactions between radio link control and TCP. Most approaches either mask adaptation in the radio link to TCP or modify TCP to reduce interactions.

The present chapter explores the possibility of relying exclusively on scheduling to reduce and possibly to eliminate the interactions for the UMTS-HSDPA systems, standardized by 3GPP for enhanced data rates UMTS operation.

In UMTS Release 99, selective ARQ is implemented in the radio network controller. It is used by RLC in acknowledgment mode. In HSDPA system, introduced in Releases 5 and 6 of UMTS, hybrid-ARQ is developed and implemented in the MAC-hs entity in the node B. As indicated previously, interaction between RLC (MAC-hs) and the TCP protocol must be analyzed to evaluate the actual performance improvements achievable by HSDPA.

Parameters that affect the TCP performance are first analyzed, and a simplified analytical model to evaluate the end-to-end TCP performance is derived. Network simulation results used to extract the TCP performance in UMTS Release 99 and HSDPA are also presented.

7.1   TCP Performance

The performance of TCP can be measured or evaluated in different ways according to the context, the system used, and the application carried over the TCP connection. In this section, TCP performance is evaluated using the following key parameters [1]:

- Effective throughput: The effective throughput, called also effective bandwidth, is the data transmission rate of the application in bits/s. The effective throughput is more significant than the communication channel throughput, since a inefficient implementation or a cross-layer interaction between TCP and RLC may result in a transmission-rate reduction at the TCP layer.
- Throughput variation: For some applications and services, it is important to know the instantaneous throughput. The throughput variation over a given timescale, depending on the application, is very important to assess end-to-end performance.
- File transfer time: The time needed to transfer the entire file. This parameter is directly related to the effective throughput.
- Round trip time: The time between the transmission of a segment and the reception of the acknowledgment. This time includes the delay in the intermediate network nodes (e.g., routers), which depends on the distance and the traffic load in the network. This parameter can limit the effective throughput.

■ Delay variation or jitter: The jitter represents the time variation in receiving packets—in other words, the variation of the RTT. This variation can have a direct impact on the appearance of triple duplicate and timeout events that result in throughput limitation and wasting of resources.

■ Fairness: An important characteristic, especially when TCP applications are carried over wireless systems. Unfair resource allocation can have drastic effects on the effective throughput. This point is addressed later in this chapter when the TCP throughput in the presence of different schedulers in the HSDPA system is evaluated.

■ Resource consumption: The amount of resources (e.g., buffer space, less transmission power) needed to achieve a given effective throughput.

7.2 General Architecture of TCP Connection over UMTS-HSDPA

Typical TCP connections between user equipment and a server via UMTS Release 99 and the HSDPA access networks are depicted in Figures 7.1 and 7.2 [2,3]. The IP datagrams are routed by the Internet to the GGSN, which connects the UMTS core network to the Internet. The data flow is forwarded from the GGSN to the user equipment using the Packet Data Protocol (PDP) used at the beginning of the packet session. Note that a PDP context contains session information such as service QoS constraints and IP addresses. Once the PDP context is established, the GGSN transmits the data to the SGSN, which relies on Iu bearers over the Iu interface to forward the data to the RNC. The data transmission between the GGSN and the RNC (UTRAN) is handled by the GPRS Tunneling Protocol (GTP-U) [3].

In the UTRAN, the data is transmitted to the user equipment using the radio access protocols—RLC and MAC entities. For example, the Packet Data Convergence Protocol (PDCP) is in charge of compressing the redundant control information of the TCP/IP protocol stack. The RLC and MAC functionalities of UMTS Release 99 and HSDPA are described in detail in Chapters 3 and 4. Note that the TCP transfer on the radio interface can be provided by one or by two peer acknowledged mode RLC entities, called radio bearers. In the case of one radio bearer, the AM RLC entity manages both the TCP data segments and the TCP feedback acknowledgments. However, in the case of two radio bearers, one AM RLC entity handles TCP data segments, and the other manages the acknowledgments [4].

The significant differences between the data transfer over UMTS Release 99 and the HSDPA systems lies in the UTRAN. In UMTS Release 99, a dedicated channel is established between the user equipment and the node B to assure the data transfer over the air interface. The node B and the

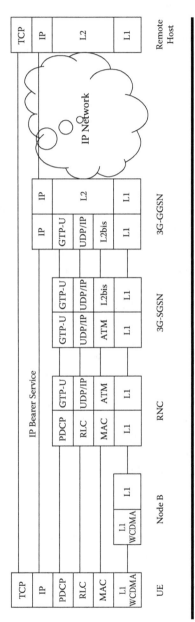

Figure 7.1 **Example of protocol stacks of a TCP connection over UMTS.**

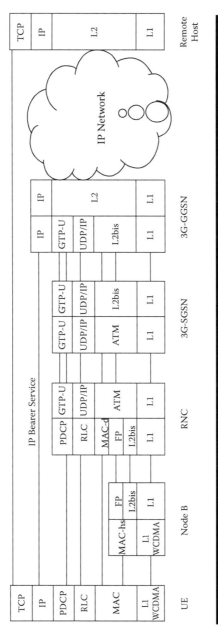

Figure 7.2 Example of protocol stacks of a TCP connection over HSDPA.

Iub interface perform a simple transfer of the data to the RNC, which controls the user equipment connection and the scheduling between different connections. In HSDPA, the scheduling is located in the node B, and one shared channel is in charge of managing different connections over the air interface. Therefore, the main data buffer is located in the node B instead of the RNC, which has a significant effect on the data transport and the TCP connection.

Losses in wireless network segments are produced by high-bit error rates and handoff procedures. As indicated in Chapters 3 and 4, the ARQ protocol is used to recover from packet losses over the radio interface, but this increases the delay of receiving packets at the TCP layer. In UMTS Release 99, the ARQ protocol is implemented in the RNC, whereas in HSDPA it is handled by the MAC-hs entity in the node B. In addition, the use of soft-combining algorithms with ARQ in the node B reduces the delay generated by the ARQ protocol. This has direct implications on the TCP connection performance and the system efficiency. Note that in both cases, Release 99 and HSDPA, the data is delivered to the TCP layer in sequence. This implies that the ARQ delay may only generate a timeout in the TCP connection. The triple-duplicate phenomenon occurs only because of Internet congestion.

When studying TCP performance in UMTS systems, either Release 99 or HSDPA, several parameters or variables can interact and affect TCP efficiency. These parameters are the

- TCP version such as Reno or SACK
- Slow-start threshold, ssthresh
- Initial congestion window, *cwnd* (1, 2 MSS or more)
- MTU size
- TCP receiver buffer size, which limits the advertised window, *awnd*
- Round-trip time in the Internet, which has a direct impact on the TCP flow rate
- Congestion rate and the segment losses in the Internet
- Error rate over the air interface and the ARQ protocol used (i.e., SR ARQ in UMTS Release 99 and N-channel HARQ SW in HSDPA)
- RLC MaxDAT, which indicates the maximal number of retransmissions of a given RLC PDU
- Allocated DCH channel (e.g., spreading factor, bit rate) in UMTS Release 99, which has a direct impact on the overall RTT value
- Scheduling algorithm used in HSDPA, which determines the transmission rate over the air interface and the delay jitter due to the variable storage duration of the data in the node B buffer
- RNC buffering capabilities in UMTS Release 99 and the node B buffering capabilities in HSDPA

- RLC transmission-window size, which indicates the maximum number of RLC PDU that can be transmitted before receiving an acknowledgment
- Number of HARQ channels in HSDPA, which indicates the number of parallel HARQ processes that can be handled by the MAC-hs entity. Any increase of this number reduces significantly the HARQ delay (as explained in Chapter 4).

7.3   Comparison among RLC, MAC-hs, and TCP

Reliable TCP protocol and the AM RLC mode (or the HARQ scheme in HSDPA) provide similar services to transmitted data such as reliability, sliding window, segmentation, and sequence numbering [4]. However, differences can be detected in the transmission-window management and the feedback acknowledgment since they are conceived for different contexts. Before studying the performance of TCP over ARQ in the UMTS system, identifying commonalities and differences in these protocols is necessary. This background should ease understanding of the TCP and HARQ interactions (for further details on this comparison see [4]).

7.3.1   Reliability

TCP, RLC, and MAC-hs rely on the ARQ scheme to recover from packet losses. All three entities ensure error-free reception of transmitted packets. The ARQ used in TCP is an adapted go-back-n protocol, where the TCP sender does not necessarily retransmit all the packets having a sequence number higher than that of the lost packet. In the enhanced TCP versions (e.g., TCP SACK explained in the previous chapter), selective acknowledgment was introduced in the TCP protocol [5–7]. In UMTS Release 99, selective ARQ is used, whereas in HSDPA an N-channel stop-and-wait scheme is implemented.

The packet loss is detected in TCP by a triple duplicate or timeout. In UMTS Release 99 and HSDPA, a negative acknowledgment is transmitted and fed back to request the retransmission of the RLC PDU, UMTS Release 99, or the MAC-hs PDU, HSDPA.

The acknowledgment in TCP is completely controlled by the receiver. The sender does not have any control to request an acknowledgment. The TCP acknowledgment is integrated in the header of a TCP segment which is ACK field over 4 bytes. In UMTS Release 99, the acknowledgment is sent, either periodically or when a RLC PDU is erroneous, by the receiver. In addition, the sender can request an acknowledgment from the receiver by setting the polling bit P contained in the RLC PDU header. The acknowledgment is transmitted through the super field (SuFi) carried in the status PDU. In HSDPA, the acknowledgment of each received MAC-hs block is transmitted over the HS-DPCCH channel and multiplexed with the channel quality indicator. The HARQ acknowledgement field is gated off when there is no ACK or NACK information being sent.

7.3.2 Flow Control and Sliding Window

TCP, MAC-hs, and RLC allow packet transmission using a sliding window so that several packets are transmitted before receiving an acknowledgment of the first packet in sequence in the sliding window.

In TCP, the transmission-window size changes dynamically according to the congestion-control algorithm. In UMTS Release 99, the initial and the maximum sizes of the transmission window are configured by the RRC entity. During the connection, the receiver can request a change of the transmission-window size. The sender informs the receiver about this change through the SuFi WINDOW field carried in a status PDU. Note that the window-size modification can be due to, for example, a limited receiver buffer size, user equipment capabilities, reception complexity, or delays. In HSDPA, the transmission-window size indicates the number of parallel HARQ processes, or instances, that can be handled by the MAC-hs entity. Note that the number of the HARQ instances is configured by the RRC entity and that up to eight HARQ channels can be used simultaneously, which limits the maximum size of the transmission window to eight MAC-hs PDUs.

7.3.3 Segmentation

RLC, MAC-hs, and TCP proceed to a segmentation to the data received from upper layers.

In TCP, the segment size is variable and is upper bounded by a maximum size called MSS. MSS is negotiated between the end TCP entities and is delimited by the size of MTU at the IP level (as explained in Chapter 5).

In UMTS Release 99, the RLC PDU size is selected during the configuration of the RLC entity. In fact, the PDU size is selected to avoid segmentation of the IP datagrams having in general a size of 40 bytes. Therefore, the PDU size is generally equal to 43 bytes (3 bytes for RLC header) except when the PDCP is used. In this case, the PDU size is adapted to the PDCP compression ratio.

In HSDPA, the RLC PDU size is selected in the same way as in UMTS Release 99. However, the MAC-hs PDU size changes dynamically according to the selected MCS. The MAC-hs PDU is transmitted over one TTI. Each RLC PDU corresponds to one or more MAC-hs PDUs.

7.4 Modeling of TCP over UMTS-HSDPA

This section presents a simple model to evaluate the performance of TCP over HSDPA. This model is an extension of the packet-loss model proposed in [8–10] (presented in 5.4.2).

The data rate at the TCP layer is computed by dividing the data size by the mean value of latency time $E(T)$; a Markov process is assumed. The mean latency time $E(T)$ is composed of T_{ss}, the latency time of the slow-start phase; T_{loss}, corresponding to the recovery time and RTO cost; and T_{ca}, representing the latency time of the steady-state phase. Hence, the data rate is given by

$$R = \frac{data}{E(T_{ss}) + E(T_{loss}) + E(T_{ca})}. \tag{7.1}$$

Consequently, modeling the effect of TCP on HSDPA requires estimates of the latency time of the slow-start phase, the loss recovery, and the steady-state phase (conducted respectively in 7.4.2, 7.4.3, and 7.4.4). The analysis of TCP timeouts needed in the latency times is presented in 7.4.1.

7.4.1 Timeout

TCP detects losses in two ways: RTOs and triple-duplicate ACKs. The RTOs of TCP can be caused by a congestion in the Internet network or by a delay due to limited bit rate or to multiple retransmissions on the radio interface generated by the ARQ technique, which increase RTT and RTOs of TCP. In this section, the probability of RTOs due to the effect of the radio interface is derived.

7.4.1.1 Proposition

The probability of RTO due to the radio interface is given by the following [11]:

$$q = Q\left(\frac{To - RTT_{wired} - \frac{1 + P_e - P_e P_s}{1 - P_e P_s} T_j}{\sqrt{\sum_m k_m \frac{W}{SF} \frac{(N \log 2(M)\tau)_{m,i}}{12000} TTI \frac{\sqrt{P_e(1 - P_e + P_e P_s)T_j}}{1 - P_e P_s}}}\right), \tag{7.2}$$

where *To* is the average duration of the first timeout in a section of one or more successive timeouts, RTT_{wired} is the average RTT of the wired part of the network, P_e is the probability of errors after decoding the information block via FEC, and P_s is the probability of errors after soft combining two successive transmissions of the same information block. K_m is the probability of selection of an MCS m. An MCS is the combination of a modulation order M, a channel coding rate τ, and a number of parallel

HSDSCH channel codes N. *TTI* is 2 ms, *SF* is 16, and W is the CDMA chip rate (3.84 Mchips/sec). Parameter T_j is the transmission time of a segment on the radio interface. This transmission time depends on the scheduler used in the MAC-hs entity to share the HS-DSCH channel among users. The analysis conducted in this section supposes that a basic fair throughput scheduler is implemented in the system. Introduction of better schedulers can only enhance performance.

7.4.1.2 Proof

In HSDPA, each TCP segment is transmitted using several predefined TTIs, each lasting 2 ms. The size of a TCP segment is 1500 octets. Transmitting a TCP segment requires between 12 and 60 TTIs, depending on the modulation and coding schemes used on the radio interface. Let S_i be the data size transmitted over each TTI. The number of retransmission required to deliver the data of size S_i is a random variable due to varying radio channel conditions. The time needed to transmit an error free TCP segment is

$$RTT = \frac{\sum_{i=1}^{n_s} N_{TTI}(i)}{n_s} T_j + RTT_{wired}. \tag{7.3}$$

Variable n_s is the number of TTIs needed to transmit a TCP segment when no errors occur on the radio interface, and $N_{TTI}(i)$ is the number of transmissions of TTI i due to HARQ. The use of scheduling on a shared channel makes the errors on each TTI independent; the successive TTIs are allocated to various users. The number of retransmission of each TTI data is independent from the other TTIs. Using the central limit theorem, the sum of a large number of independent and identically distributed (iid) symmetric random variables can be considered as a Gaussian variable. Hence, the number of transmissions of a TCP segment $N_i = \frac{\sum_{i=1}^{n_s} N_{TTI}(i)}{n_s}$ can be modeled by a Gaussian variable. Consequently, the time needed to transmit a TCP segment (RTT) is a Gaussian variable. The probability of timeout RTO expressed as $prob(RTT = \text{Gaussian} > To)$ with the Gaussian assumption leads to $Q(\frac{To - E(RTT)}{\sigma(RTT)})$. By evaluating and replacing $E(N_i)$ and $\sigma(N_i)$ by their values, $E(RTT)$ and $\sigma(RTT)$ are obtained, and the probability of RTO has the form provided previously in (7.2). A detailed derivation of this proof is provided in [11,12].

7.4.2 Slow Start*

The TCP connection begins in slow-start mode where it quickly increases its congestion window to achieve best-effort service until it detects a packet loss.

---

* From [13], page 378. With kind permission of Springer Science and Business Media.

In the slow-start phase, the window size *cwnd* is limited by a maximum value W_{max} imposed by the sender or receiver buffer limitations. To determine $E(T_{ss})$, the number of data segments $E(d_{ss})$ the sender is expected to send before losing a segment is needed. From this number, one can deduce $E(W_{ss})$, the window one would expect TCP to achieve at the end of the slow start where there is no maximum window constraint. If $E(W_{ss}) \leq W_{max}$, then the window limitation has no effect, and $E(T_{ss})$ is simply the time for a sender to send $E(d_{ss})$ in the exponential growth mode of the slow start. On the other hand, if $E(W_{ss}) > W_{max}$ then $E(T_{ss})$ is the time for a sender to slow start up to $cwnd = W_{max}$ and then to send the remaining data segments at a rate of W_{max} segments per round.

Let e be the probability of retransmission (congestion + RTO). Probability e can be evaluated using the following equation:

$$e = p + q - pq. \tag{7.4}$$

$E(d_{ss})$ can be calculated using the following expression:

$$E(d_{ss}) = \left(\sum_{k=0}^{d-1} (1-e)^k e.k \right) + (1-e)^d . d$$

$$= \frac{(1-(1-e)^d)(1-e)}{e}, \tag{7.5}$$

where d is the number of segments in the file. Using the same demonstration as in [10], the mean value of the latency time can be evaluated as follows:

$$E(T_{ss}) = \begin{cases} RTT \left[log_\gamma \left(\frac{W_{max}}{W_1} \right) + 1 + \frac{1}{W_{max}} (E(d_{ss})) - \frac{\gamma W_{max} - W_1}{\gamma - 1} \right) \right] \\ \qquad\qquad\qquad\qquad \text{When } E(W_{ss}) > W_{max} \\[1em] RTT . log_\gamma \left(\frac{E(d_{ss})(\gamma - 1)}{W_1} + 1 \right) \\ \qquad\qquad\qquad\qquad \text{When } E(W_{ss}) \leq W_{max} . \end{cases} \tag{7.6}$$

γ is the rate of exponential growth of the window size during the slow start. $E(W_{ss})$ is given by

$$E(W_{ss}) = \frac{E(d_{ss})(\gamma - 1)}{\gamma} + \frac{W_1}{\gamma}. \tag{7.7}$$

7.4.3 Recovery Time of the First Loss*

The slow-start phase in TCP ends with the detection of a packet loss. The sender detects a loss in two ways: negative ACK (triple duplicate) or RTOs. The RTO could be caused by a congestion in the wired network or by the retransmissions on the radio interface. After an RTO, the window size decreases to one; however, the loss detected by the triple-duplicate ACKs decreases the window size to one-half. This section evaluates the recovery time of this first loss. The probability of loss in a file of d TCP segments is

$$loss = 1 - (1 - e)^d. \tag{7.8}$$

The loss between segments could be considered independent [8,9]. Let $Q'(e, w)$ be the probability that if a loss occurs it is an RTO. This probability can be evaluated as follows: Let *cong* and *wirel* be, respectively, the probabilities that there is a congestion loss in the transmission of the file and there is an RTO due to radio interface conditions.

$$cong = 1 - (1 - p)^d \tag{7.9}$$

$$Wirel = 1 - (1 - q)^d, \tag{7.10}$$

where q is evaluated in 7.4.1 by (7.2). [8,9] derive the probability that a sender in congestion avoidance will detect a packet loss with an RTO, as a function of congestion rate p and window size w. This probability is denoted by $F(p, W)$:

$$F(p, W) = min\left(1, \frac{(1 + (1 - p)^3(1 - (1 - p)^{w-3}))}{(1 - (1 - p)^w)/(1 - (1 - p)^3)}\right). \tag{7.11}$$

The probability of RTO is simply obtained through

$$RTO = cong.F(p, W) + Wirel - Wirel.cong.F(p, W). \tag{7.12}$$

Hence, the probability $Q'(e, w)$ is derived as [12,13]

$$Q'(e, w) = \frac{RTO}{loss}$$

$$= \frac{cong.F(p, W) + Wirel - Wirel.cong.F(p, W)}{1 - (1 - e)^d} \tag{7.13}$$

The probability of loss via triple duplicate is $loss(1 - Q'(e, w))$. It is assumed that fast recovery for a triple duplicate takes one RTT [10]. However, it takes

---
* From [13], page 379. With kind permission of Springer Science and Business Media.

more time for an RTO. The RTO cost, derived in [8,9], does not take into account the radio interface effects. Using the mean expected cost of an RTO

$$E(z^{TO}) = \frac{1 + e + 2e^2 + 4e^3 + 8e^4 + 16e^5 + 32e^6}{1 - e} T_O \qquad (7.14)$$

and combining these results, the mean recovery time at the end of the initial slow start is obtained [12-14]:

$$E(T_{loss}) = loss \left(Q'(e, w)E(z^{TO}) + (1 - Q'(e, w))RTT \right). \qquad (7.15)$$

7.4.4   Steady-State Phase*

The time needed to transfer the remaining data can be derived in the same way as in [10]. Indeed, the amount of data left after the slow start and any following loss recovery is approximately

$$E(d_{ca}) = d - E(d_{ss}). \qquad (7.16)$$

This amount of data is transferred with a throughput $R(e, RTT, To, W_{max})$. The latency time is then given by

$$E(T_{ca}) = \frac{E(d_{ca})}{R(e, RTT, To, W_{max})}. \qquad (7.17)$$

In [8,9], the throughput $R(p, RTT, To, W_{max})$ is evaluated without the radio interface effects. By using the same demonstration as in [8,9] and by introducing e, RTT and $Q'(e, w)$ provided in this section, the derivation of the throughput expression leads to the following equation:

$R(e, RTT, To, W_{max})$

$$= \begin{cases} \dfrac{\frac{1-e}{e} + \frac{W(e)}{2} + Q'(e, W(e))}{RTT\left(\frac{b}{2}W(e) + 1\right) + \frac{Q'(e, W(e))G(e)To}{1-e}} & \text{When } W(e) < W_{max} \\[4ex] \dfrac{\frac{1-e}{e} + \frac{W_{max}}{2} + Q'(e, W_{max})}{RTT\left(\frac{b}{8}W_{max} + 2 + \frac{1-e}{eW_{max}}\right) + \frac{Q'(e, W_{max})G(e)To}{1-e}} & \text{When } W(e) \geq W_{max}, \end{cases} \qquad (7.18)$$

where b is the number of TCP segments acknowledged by one ACK and $W(e)$ is given by

$$W(e) = \frac{2+b}{3b} + \sqrt{\frac{8(1-e)}{3be} + \left(\frac{2+b}{3b}\right)^2}. \qquad (7.19)$$

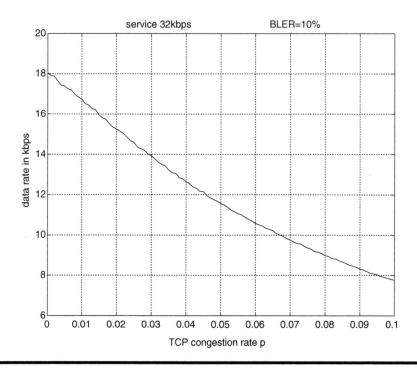

Figure 7.3 Effect of TCP on 32 kbps applications performance. (From [13], page 386. With kind permission of Springer Science and Business Media.)

Finally, once the total latency time is calculated, the bit rate for each service at the TCP layer can be evaluated using equation (7.1). Figures 7.3, 7.4, and 7.5 present the TCP throughput variations according to the congestion rate in the wired network, respectively, for 32, 64, and 128 Kbps.

7.4.5 Effect of TCP on Wireless Network

The decrease of the TCP bit rate over the radio interface is due to two reasons: decrease of TCP window size, and retransmissions of TCP segments. In the case of dedicated channels [3,4,15–25], it is interesting to evaluate the final TCP bit rate since the number of users is fixed. However, when several users share the same channel in time, the performance of TCP includes the user bit rate and the system throughput (as explained in 6.2.1—scheduling over reliable shared channel). The evaluation of the mean number of TCP segments retransmissions N_{TCP} becomes important. When N_{TCP} has a low value compared to the decrease of TCP window size (see Figure 7.6), the decrease of TCP bit rate is essentially due to the decrease of window size. In this case, the number of TCP packets arriving at the node B decreases, and more TTIs are available on the shared channel. By allocating these

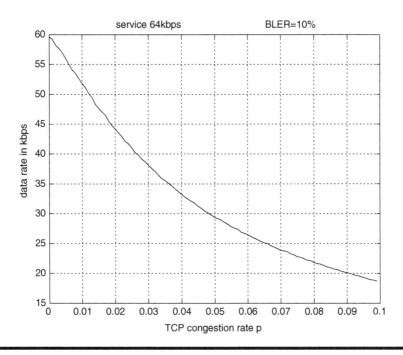

Figure 7.4 Effect of TCP on 64 kbps applications performance. (From [13], page 387. With kind permission of Springer Science and Business Media.)

TTIs to the other users, the radio interface bit rate can be increased: the transmission rate of each TCP segment, which limits the increase of RTT and consequently reduces the degradation of TCP bit rate.

N_{TCP} can be evaluated using the probability of transmitting segments n times before correct reception. The probability that a segment is transmitted only once is $(1 - e)$. The TCP segment is transmitted two times with a probability $e(1 - e)$. The retransmission of a segment could be caused by an RTO or a triple duplicate. In the case of a RTO, the timeout (TO) period is *To*. If another timeout occurs, *To* doubles to 2 *To*. This doubling is repeated for each unsuccessful retransmission until a *To* of 64 *To* is reached, after which the *To* is kept constant at 64 *To*. However, in the case of a triple duplicate, a timeout still equals *To*. When the window is small, the retransmission is due to a RTO. Practically in a wireless network, after the third retransmission the loss is only due to the timeout *To*. The probability of RTO due to the radio interface is q as given by (7.2). If the time out is 2 *To*, in the same way the probability of a RTO is defined as $q2$ (by replacing *To* by 2 *To*). Probabilities $q4, q8, \ldots q64$ are defined similarly according to number of retransmissions. Let

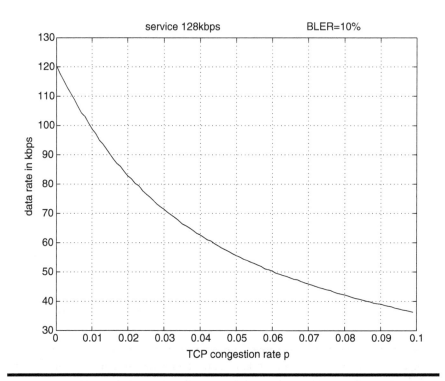

Figure 7.5 Effect of TCP on 128 kbps applications performance. (From [13], page 387. With kind permission of Springer Science and Business Media.)

$x2 = (1 - p)(1 - q2)$ and define $x4, x8, \ldots x64$ on the same basis. The probability to have three transmissions is consequently equal to $ex2(pF(p, W) + q - pF(p, W)q) + e(1-e)p(1 - F(p, W))$. The probability to have 4 transmissions is $e^2(1 - x2)x4$. The TCP segment is transmitted 5 times with a probability of $e^2(1 - x2)(1 - x4)x8$. The ensuing probabilities have the same form. (From [14], page 4107, ©2004 IEEE.)

Based on the expressions of these probabilities and following several manipulations, the mean number of transmissions N_{TCP} is reported in (7.20) [12–14].

$$N_{TCP} = 1 + e - 2e^2 + 3ex2(pF(p, W) + q - pF(p, W)q)$$

$$+ 3e(1 - e)p(1 - F(p, W)) + 4e^2(1 - x2)x4$$

$$+ 5e^2(1 - x2)(1 - x4)x8 + 6e^2(1 - x2)(1 - x4)(1 - x8)x16$$

$$+ e^2(1 - x2)(1 - x4)(1 - x8)(1 - x16)\frac{(1 + 7x16)}{x16}. \qquad (7.20)$$

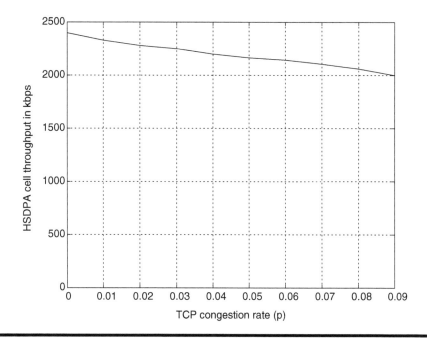

Figure 7.6 **Effect of TCP on HSDPA cell capacity when the proportional fair scheduler is used.**

N_{TCP} can then be introduced in the cell throughput analytical model to determine the interaction between TCP and HS-DSCH. This has been conducted in [11] for several schedulers. In this section, the cell throughput variation according to the congestion rate is depicted in Figure 7.6 where a fair throughput scheduler is assumed on the HS-DSCH.

To mask the TCP degradation in HSDPA system, a refined proportional fair algorithm was proposed in [26] (similar approach is proposed independently in [11], see Section 6.2.1.4). The TCP user throughput according to the number of cell users, when the TCP congestion rate is equal to 3 percent (an acceptable mean value [25]), is depicted in Figure 7.7. For the same number of users, the refined PF gives better results than the PF based on the radio interface only. The gain is about 30 percent for six users and 35 percent for 12 users. Further analysis and results on TCP in HSDPA system for several schedulers are provided in [11].

7.5 Other Analyses of TCP over UMTS-HSDPA

Additional studies that focused only on the interaction between TCP and the variable delay caused by ARQ over wireless channels in 3G system are available in [15–25]. These studies propose enhancements in the TCP

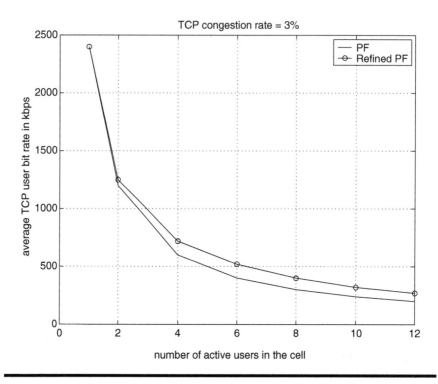

Figure 7.7 TCP user bit rate according to the number of users in the cell for pro-portional fair and modified proportional fair.

behavior, whereas others consider the 3G system (UMTS) in all its aspects and try to evaluate the effect of the overall system parameters and layers (e.g., MAC, RLC) on the TCP throughput degradation [3,4,27–32]. For example, the effect of the TCP slow start (initial congestion-window size) on the TCP performance over a DCH in UMTS can be found in [3]. The effect of RLC buffer size, RLC MaxDAT, receiver buffer size, congestion rate, RTT in the Internet, RLC transmission window, and the CDMA code tree utilization (i.e., the percentage of code tree allocated to TCP services, or the fraction of cell capacity assigned to TCP services) on the TCP throughput in UMTS system are provided in [3,4,27–30].

Note that all these studies consider only the interaction of TCP with dedicated UMTS channels. For the HSDPA system, fewer studies are available in the literature [3,4,11–13,33–35]. Recently, researchers started investigating the impact of the variable delay introduced by the schedulers in HSDPA on the TCP system performance. [33, 34] propose an interesting approach that improves the long-lived TCP performance while reducing the latency of short TCP flows. These methods achieve better wireless channel utilization by simultaneously using two algorithms: a network-based solution called window regulator and a PF scheduler with rate priority (PF-RP). The window-regulator algorithm "uses the receiver window field in the acknowledgment packets to convey the instantaneous wireless channel conditions to the TCP source and an ACK buffer to absorb the channel variations," as stated in [33, pg. 1893]. This approach is found to increase TCP performance for any given buffer size. The PF-RP scheduler part of the proposal, based on the simultaneous use of two schedulers PF and PF with strict priority (PF-SP), tries to differentiate short flows from long flows by assigning different priorities to the flows. This scheduler achieves a trade-off between fairness among users and system throughput maximization and also minimizes short-flow latency. Results in [32,34] indicate that combining the window regulator and the PF-RP scheduler "improves the performance of TCP Sack by up to 100% over a simple drop-tail algorithm for small buffer sizes at the congested router."

Authors in [35] propose an opportunistic scheduler that reduces packet delay in the node B buffer, achieves better fairness between users, and improves TCP performance without losing cell capacity. This scheduler has the advantage of also being suitable for streaming services.

These studies (including the model in 7.4) on the impact of variable delays, caused by schedulers over time-shared channels using ARQ for reliable transmission on TCP performance appear as a promising approach to effectively reduce the TCP throughput degradation in wireless systems. These types of analyses should be pursued to exploit even further user diversity (users experience different random short-term channel variations), to maximize the overall system throughput, to reduce TCP performance degradations, and to achieve fairness among TCP flows. Accomplishing this joint optimization remains a challenge that requires an extended definition of metrics to achieve overall trade-off in terms of efficiency and fairness.

References

1. Hassan, Mahbub, and Raj Jain 2004. *High Performance TCP/IP Networking*. Upper Saddle River, NJ: Prentice Hall.
2. 3GPP TS 23.060 V6.10.0. 2005. Service Description. Stage 2, Release 6, September.
3. Ameigeiras Gutierrez, P. J. 2003. Packet Scheduling and QoS in HSDPA. PhD diss. Aalborg University, October.
4. Bestak, R. 2003. Reliability Mechanisms (Protocols ARQ) and Their Adaptation in 3G Mobile Networks. PhD diss. ENST Paris, December 18, http://pastel.paristech.org/archive/00000514/01/RBestakThese.pdf.
5. Mathis, M., J. Mahdavi, S. Floyd, and A. Romanow. 1996. Selective Acknowledgment Option, RFC-2018, October.
6. Floyd, S., J. Mahdavi, M. Mathis, and M. Podolsky. 2000. An Extension to the Selective Acknowledgement (SACK) Option for TCP. RFC 2883, July.

7. Mathis, M., and J. Mahdavi. 1996. Forward Acknowledgment: Refining TCP Congestion Control. In Proc. of the ACM Annual Conference of the Special Interest Group on Data Communication (SIGCOMM 96), (August).

8. Padhye, J., V. Frnoiu, D. Towsley, and J. Kurose. 1998. Modeling TCP Throughput: A Simple Model and Its Empirical Validation. Proc. of ACM Annual Conference of the Special Interest Group on Data Communication (SIGCOMM).

9. Padhye, J., V. Frnoiu, D. Towsley, and J. Kurose. 2000. Modeling TCP Throughput: A Simple Model and Its Empirical Validation. *IEEE/ACM Transactions on Networking* 8, no. 2:133–45 (April).

10. Cardwell, N., S. Savage, and T. Anderson. 2000. Modeling TCP Latency. *Paper presented at the IEEE Conference on Computer Communication (IN-FOCOM)* 3:1742–51 (March 26–30).

11. Assaad, Mohamad, and Djamal Zeghlache. 2006. Cross Layer Design in HSDPA System, *IEEE Journal on Selected Areas in Communications*, forthcoming.

12. Assaad, Mohamad, and Djamal Zeghlache. 2005. How to Minimize the TCP Effect in a UMTS-HSDPA System. *Wiley Wireless Communications and Mobile Computing Journal* 5, no. 4:473-485 (June).

13. Assaad, Mohamad, Badii Jouaber, and Djamal Zeghlache. 2004. TCP Performance over UMTS-HSDPA System. *Kluwer on Telecommunication Systems Journal* 27:2–4, 371–391.

14. Assaad, M., B. Jouaber, and D. Zeghlache. 2004. Effect of TCP on UMTS/HSDPA System Performance and Capacity. *IEEE Global Telecommunications Conference, GLOBECOM04*, Dallas. 6:4104–4108.

15. Ludwig, R., B. Rathonyi, A. Konrad and A. Joseph. 1999. Multilayer Tracing of TCP over a Reliable Wireless Link. ACM SIGMETRICS International Conference on Measurement and Modeling of Computer systems, May.

16. Yavuz, M., and F. Khafizov. 2002. TCP over Wireless Links with Variable Bandwith. Proc. of IEEE VTC (September).

17. Xylomenos, G., G. C. Polyzos, P. Mahonen, and M. Saaranen. 2001. TCP Performance Issues over Wireless Links. *IEEE Communication Magazine* 39, no. 5:52–8 (April).

18. Balakrishnan, H., V. Padmanabhan, S. Seshan, M. Stemm, and R. H. Katz. 1997. A Comparison of Mechanisms for Improving TCP Performance. *IEEE/ACM Transactions on Networking* 5:756–69 (December).

19. Chaskar, N. M., T. V. Lakshman, and U. Madhow. 1999. TCP over Wireless with Link Level Error Control: Analysis and Design Methodology. *IEEE/ACM Trans. Networking* 7:605–15 (October).

20. Wong, J. W. K., and V. C. M. Leung. 1999. Improving End to End Performance of TCP Using Link Layer Retransmissions over Mobile Internetworks. In Proc. of the IEEE International Conference on Communication (ICC), 324–8.

21. Bai, Y., A. T. Ogielski, and G. Wu. 1999. Interaction of TCP and Radio Link ARQ Protocol. In proc. of the IEEE VTC, 1710–14.

22. Wang, K. Y., and S. K. Tripathi. 1998. Mobile End Transport Protocol: An Alternative to TCP/IP over Wireless Links. In Proc. of the IEEE Conference on Computer Communication (INFOCOM), 1046–53.

23. Hossain, E., D. I. Kim, and V. K. Bhargava. 2004. Analysis of TCP Performance under Joint Rate and Power Adaptation in Multicell Multirate WCDMA Packet Data Systems. *IEEE Transactions on Wireless Communications* 3, no. 3:865–79 (May).

24. Peisa, J., and E. Englund. 2002. TCP Performance over HS-DSCH, In Proc. of the IEEE Vehicular Technology Conference (VTC)'02. 2:987–91 (May).

25. Meyer, M. 2001. Analytical Model for TCP File Transfers over UMTS. Paper presented at the 3GWireless Conference, San Francisco, May-June.

26. Klein, T. E., K. K. Leung, and Haitao Zheng. 2004. Improved TCP Performance in Wireless IP Networks through Enhanced Opportunistic Scheduling Algorithms. *IEEE GLOBECOM* 5:2744–8 (November 29–December 3).

27. Gurtov, Andrei. 2002. Efficient Transport in 2.5G 3G Wireless Wide Area Networks. PhLic diss. C-2002-42, Department of Computer Science, University of Helsinki, September.

28. Gurtov, Andrei. 2000. TCP Performance in the Presence of Congestion and Corruption Losses. Master's thesis, C-2000-67, Department of Computer Science, University of Helsinki, December.

29. Kuhlberg, Panu. 2001. Effect of Delays and Packet Drops on TCP-Based Wireless Data Communication. Master's thesis, C-2001-7, Department of Computer Science, University of Helsinki, February.

30. Sarolahti, Pasi. 2001. Performance Analysis of TCP Enhancements for Congested Reliable Wireless Links. Master's thesis C-2001-8, Department of Computer Science, University of Helsinki, February.

31. Kulve, Tuomas. 2003. Analysis of Concurrent TCP and Streaming Traffic over a Wireless Link. Master's thesis C-2003-50, Department of Computer Science, University of Helsinki, October.

32. Saarto, Jarno. 2003. WWW Traffic Performance in Wireless Environment. Master's thesis C-2003-35, Department of Computer Science, University of Helsinki, May.

33. Choon Chan, Mun, and Ram Ramjee. 2004. Improving TCP/IP Performance over Third Generation Wireless Networks. Paper presented at the IEEE Conference on Computer Communication (INFOCOM)04, Hong Kong, March, 1893-1904.

34. Choon Chan, Mun, and Ram Ramjee. 2002. TCP/IP Performance over 3G Wireless Links with Rate and Delay Variation. Paper presented at ACM Conference on Mobile Computing and Netuoring (MOBICOM)02, Atlanta, GA, September.

35. Assaad, Mohamad, and Djamal Zeghlache. Opportunistic Scheduler for HSDPA System. *IEEE Transactions on Wireless Communications*, forthcoming.

Index

Milton Keynes UK
Ingram Content Group UK Ltd.
UKHW040102071024
449327UK00019B/738